CHAPTER ONE

The Basics

1A   Introduction and Apology

Since the appearance about 2300 years ago of Euclid's book, *The Elements*, it has been traditional to study geometry as a *deductive* discipline, using the so-called axiomatic method. Ideally, this means that the geometer (or geometry student) starts with a few assumed facts called axioms or postulates and then systematically and carefully derives the entire content of the subject using nothing but pure logic. The axiomatic method requires that this content, which for geometry is a collection of facts about triangles, circles, and other figures, should be presented as a sequence of ever deeper theorems, each rigorously proved using the axioms and earlier theorems.

Geometry has continued to develop in the centuries since Euclid. Mathematicians, including many talented amateurs, have discovered a wealth of beautiful, surprising, and even spectacular properties of geometric figures, and these, together with the geometry of *The Elements*, constitute what is today called Euclidean geometry. Amazing as many of these asserted facts may be, we can be confident that they are correct because they are *proved*; they are not merely tested by experiment and confirmed by measurement and observation. Furthermore, we do not have to rely on the authority of Euclid or his successors, because with a little practice and preparation, we can read and understand these proofs ourselves. With luck, we might find simpler and more elegant proofs for some of the known theorems of Euclidean geometry. It is even possible that we might discover new geometric facts that have never been seen before and invent our own proofs for them.

The development of geometry has resulted not only in the discovery of amazing facts, but also in the invention of new and powerful techniques of proof, and these too are considered part of Euclidean geometry. Descartes' invention of coordinate geometry is an example of this. A third thread in the history of geometry, especially in modern times, has been an investigation into the foundations of the subject. It has been asked, for instance, whether or not all of Euclid's axioms and postulates are really valid and whether or not we really need to assume them all. But we say almost nothing about these foundational issues in this book. Instead, we concentrate on the two more classical themes of geometry: facts and proofs. We offer the reader a selection of some of the

1

more striking facts of Euclidean geometry, and we present enough proof machinery to establish them.

In the centuries since Euclid, the accepted standards of mathematical rigor have changed, and the sufficiency of proofs in the style of Euclid has been challenged. One objection, for example, is that Euclid relied too heavily on diagrams and that he and the other classical geometers did not always prove facts they considered to be obvious from the figures. In this book, we follow the tradition of Euclid and of most of his successors, and so we are willing to use some of the information contained in carefully drawn figures. But we must not rely on diagrams for certain other types of information, and unfortunately, it is difficult to be precise about exactly what we are willing to read off from a diagram and what requires a proof. We are confident that readers will catch on to this fairly quickly, however, and we apologize for the ambiguity. Perhaps an example will help clarify the situation.

In Figure 1.1, angle bisectors AX, BY, and CZ have been drawn in $\triangle ABC$. Three facts that seem clear in the figure are:

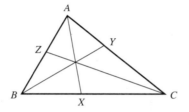

Figure 1.1

1. $BX < XC$. In other words, line segment BX is shorter than line segment XC.
2. The point X, where the bisector of $\angle A$ meets line BC, lies on the line segment BC.
3. The three angle bisectors are concurrent.

Before we discuss these observations, let us be clear about the meanings of the technical words and notation. We assume that the nouns *point*, *line*, *angle*, *triangle*, and *bisector* are familiar to readers of this book. Two lines, unless they happen to be parallel, always meet at a point, but if three or more lines all go through a common point, then something unexpected is happening and we say that the lines are **concurrent**. The remaining undefined technical word in the preceding list of facts is *segment*. A line **segment** is that part of a line that lies between two given points on the line. Observe that in assertion (1), we are using the notation BX in two different ways: In the inequality, BX represents the *length* of the line segment, which is a number, and then later, BX is used to name the segment itself. In addition, the notation BX is often used to represent the entire line containing the points B and X, and not just the segment they determine. But we shall always provide enough information so that it is clear from context which of the three possible meanings is intended. For example, in inequalities or equations, it is obviously the numerical interpretation that is wanted.

Each of (1), (2), and (3) is true, but we need to distinguish among three different types of truth here. First, note that the fact in (1) that $BX < XC$ is an accident. This

inequality happens to be true in this figure, but it is not an instance of some general or universal truth. Even in a particular case, this sort of information can be unreliable since it depends on the accuracy of the diagram and the care with which measurements are made. It is never considered valid to read off an equality from a diagram, and it is rare that one is justified in reading off an inequality.

Fact (2), that point X lies between points B and C on line BC, is not accidental. Indeed, it seems completely obvious that the bisector of each angle of an arbitrary triangle must always intersect the opposite side of the triangle. In other words, the bisector intersects the line *segment* determined by the other two vertices of the triangle. Note that although this is true about angle bisectors, it can fail for other important lines associated with a triangle. For example, the **altitude** drawn from A, which is the line through A perpendicular to line BC, may not meet the segment BC. Although this failure (accidentally) does not happen for $\triangle ABC$ of Figure 1.1, it certainly can happen for other triangles. The "obvious" fact that angle bisectors of triangles meet the opposite sides is the sort of information that Euclid was, and we are, willing to read off from diagrams.

Although it is a fact that for every triangle, the three angle bisectors are concurrent, as in assertion (3), we follow Euclid in that we do not consider this to be obvious. The experimental evidence of even an accurate computer-drawn diagram (or of many such diagrams) is not sufficient for us to accept this as a universal truth; we require a proof. Readers will probably have seen a proof of this theorem in high-school geometry, and it is also proved here in Chapter 2.

Thus, there is an inherent ambiguity about which information can be reliably established from diagrams and which cannot. Because of this ambiguity, modern standards of mathematical rigor require that there must be no reliance whatsoever on figures. Geometry without diagrams was made possible by David Hilbert (1862–1943) who, building on the work of his predecessors, constructed an appropriate set of precisely stated axioms from which he was able to prove everything formally and without diagrams. Euclid's ideal that geometry should be a purely deductive enterprise was thus finally realized by Hilbert at the turn of the 20th century.

In particular, Hilbert's axioms allowed him to prove our observation (2), that an angle bisector of a triangle always meets the opposite side of the triangle. But it is far from a triviality to prove rigorously, and without relying on a diagram, that the bisector of $\angle BAC$ meets line BC at a point X that lies between B and C. Furthermore, there would be no hope of proving such a thing without having a precise definition of the word *between*, which Hilbert was able to provide.

The achievements of Hilbert and other researchers into the foundations of geometry were substantial and significant, but it is my opinion that, by far, the most interesting theorems of geometry are those that provide surprises. We feel, for example, that the fact that the three angle bisectors of a triangle are always concurrent is much more exciting than is the fact that the angle bisectors meet the opposite sides of the triangle. For this reason, we have chosen to present in this book as many unexpected facts and surprising theorems as space allows. Since we want to get to these quickly, we must omit Hilbert's careful and rigorous treatment of the foundations of geometry, and instead, we will follow tradition and rely on diagrams without specifying exactly the extent of our

reliance. Also, since the readers of this book will have studied some geometry in high school, we will not start our presentation at the very beginning of the subject.

We have already apologized for the ambiguity about how much information we are allowed to obtain from diagrams. Some apology is also appropriate concerning the issue of how much high-school geometry we are assuming. Students may fairly complain when they are doing homework exercises that it is unclear which facts they must prove and which they can merely quote as remembered from school. We do not attempt to give a complete list of assumed results, but we shall show by example the level of proof that we expect, and we shall devote much of the rest of this chapter to a review of some of the essential facts, definitions, and theorems from high-school geometry.

Before proceeding with our review of high-school geometry, we discuss briefly some of the issues in the foundations of geometry to which we referred earlier. We mentioned that Euclid called his unproved assumptions axioms and postulates. The distinction, which is not considered significant today, is that Euclid's axioms concerned general logical reasoning, while his postulates were more specifically geometric. For example, one of Euclid's axioms is "things equal to the same thing are equal to each other," whereas his parallel postulate essentially asserts that "given a line and a point not on that line, there exists one and only one line through the given point parallel to the given line." Actually, Euclid's parallel postulate is not stated in precisely this way; it appears in a somewhat more complicated but logically equivalent formulation.

Over the centuries, Euclid's parallel postulate has engendered a great deal of interest and controversy. Somehow, the existence and uniqueness of a line parallel to a given line through a given point seemed less obvious than the facts asserted by the other postulates. Geometers felt uncomfortable assuming the parallel postulate, and they attempted instead to prove it; they tried to deduce it from the rest of Euclid's axioms and postulates. Before we discuss these attempts, we should stress that the parallel postulate makes two separate assertions, each of which would have to be proved. It would be necessary to show that a parallel line (through the given point) *exists*, and one would also need to prove that it is *unique*.

How might a proof of the parallel postulate proceed? One could assume that it is false and then try to derive some contradictory conclusions. Assume, for example, that there is some line AB and some point P not on AB, and there is no line parallel to AB through P. If by means of this assumption one could deduce the existence of a triangle $\triangle XYZ$ for which $XY > YZ$ and also $XY < YZ$, then this contradiction would prove at least the existence part of the parallel postulate. What actually happened when this was tried was that *apparent* contradictions were derived and the existence of figures that *seem* impossible was proved. For example, the assumption that no line parallel to AB goes through point P yields a triangle $\triangle XYZ$ that has three right angles. "Surely this is impossible," we might say, but how can we prove this impossibility? We know from high school that the three angles of a triangle sum to 180°, and an easy experiment confirms this fact. (Tear off the three corners of a paper triangle and line them up to see that the three angles total a straight angle.) It may seem that by proving the existence of a triangle with three right angles, we have the desired contradiction, but this is wrong. The high-school proof that $\angle X + \angle Y + \angle Z = 180°$ ultimately relies on the parallel postulate, which we are temporarily refusing to assume. Also, the experiment with paper triangles is certainly not a mathematical proof.

After repeated attempts to obtain contradictions from the denial of either the existence or the uniqueness part of the parallel postulate, it was realized by J. Bolyai, N. Lobachevski, and C. Gauss in the 19th century that, although such denials yield seemingly ridiculous situations, such as triangles with three right angles, no proof of a contradiction was possible. In fact, it was *proved* that no such proof is possible. In other words, one could build a perfectly consistent deductive geometry by replacing Euclid's parallel postulate with either one of two alternative new postulates. One of these denies the existence of a parallel to some line through some point, and the other asserts the existence of at least two such parallels. Each of the two types of geometry that arise in this way is said to be non-Euclidean, and each has its own set of proved theorems. The geometry where no parallel exists is called elliptic geometry, and when more than one parallel to a line goes through a point, we have hyperbolic geometry. The deductions of each of the two types of non-Euclidean geometry contradict each other, and they also contradict the theorems of classical Euclidean geometry, but each of the three flavors of geometry appears to be internally consistent, and from a mathematician's point of view, they are equally valid. Actually, it is known that if Euclidean geometry is internally consistent, then the two non-Euclidean geometries are also consistent, but no formal proof of the consistency of Euclidean geometry has been found.

The question of which, if any, of the three flavors of geometry that we have been discussing describes the real world is interesting, but it is not really a mathematical question; it belongs in the realm of physics. The little experiment with the paper triangle certainly suggests that we live in a Euclidean universe, but this has not been firmly established on a large scale. Indeed, modern theories of the structure of the cosmos, including Einstein's theory of general relativity, suggest that none of the three geometries provides an entirely accurate description of the universe in which we actually live. Nevertheless, Euclidean geometry provides a very good approximation to reality on a human scale, and so it is useful for practical purposes such as engineering, navigation, and architecture.

1B   Congruent Triangles

We devote the rest of this chapter to some of the basic facts and techniques of Euclidean geometry, much of which will be a review for most readers. Recall that two geometric figures are **congruent** if, informally speaking, they have the same size and shape. Somewhat more precisely, two figures are congruent if one can be subjected to a rigid motion so as to make it coincide with the other. By a rigid motion, we mean a translation or shift, a rotation in the plane or a reflection in a line. The latter can be viewed informally as lifting the figure from the plane, flipping it over, and placing it back in the plane.

In Figure 1.2, for example, the three triangles are congruent, although to make $\triangle XYZ$ coincide with either of the other two triangles it is necessary to reflect it or flip it over. We write $\triangle ABC \cong \triangle RST$ to report that the first two triangles in Figure 1.2 are congruent, but note that there is more to this notation than may at first be apparent. The only way that these two triangles can be made to coincide is for point R to coincide with point A, for S to coincide with B, and for T to coincide with C. We say that A and R, B and S, and C and T are **corresponding points** of these two congruent triangles. The only

 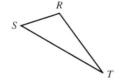

Figure 1.2

correct ways to report the congruence is to list corresponding points in corresponding positions. It is correct, therefore, to write $\triangle ABC \cong \triangle RST$ or $\triangle BAC \cong \triangle SRT$, but it is wrong to write $\triangle ABC \cong \triangle SRT$. This latter assertion is not true because there is no way that these two triangles in Figure 1.2 can be made to coincide with A and S, B, and R, and C and T being corresponding points.

Since $\triangle RST \cong \triangle XYZ$, it is clear that corresponding sides of these triangles have equal length and that corresponding angles have equal measure (contain equal numbers of degrees or radians). We can thus write, for example, $RS = XY$ and $\angle SRT = \angle YXZ$. Note that in this context, the notation $\angle SRT$ refers to the measure of the angle in some convenient units, such as degrees or radians. In other situations, however, we may write $\angle SRT$ to refer to the angle itself. This is entirely analogous to the fact that RS can refer either to a line segment or to its length in centimeters, inches, miles, or whatever. In some geometry books, the notation $m\angle SRT$ is used to denote the measure of $\angle SRT$.

Whenever we know that two triangles are congruent, we can deduce six equalities: three of lengths and three of measures of angles. It is also reasonably obvious that given two triangles, if all six equalities hold, then the triangles can be made to "fit," one on top of the other, and they are congruent. As readers of this book are surely aware, it is not necessary to know all six equalities to conclude that two triangles are congruent. If we know, for example, that the three sides of one triangle equal, respectively, the three corresponding sides of the other triangle, we can safely deduce that the triangles are congruent. If we know, for example, that $AB = RS$, $AC = RT$, and $BC = ST$, we can conclude that $\triangle ABC \cong \triangle RST$. In a proof, where each assertion must be justified, we say that these two triangles are "congruent by SSS." The abbreviation SSS, which stands for "side-side-side," refers to the theorem that says that if the three sides of a triangle are equal in length to the three corresponding sides of some other triangle, then the two triangles must be congruent.

Other valid criteria that can be used to prove that two triangles are congruent are SAS, ASA, and SAA. These, of course, are abbreviations for "side-angle-side," "angle-side-angle," and "side-angle-side," respectively. In the expectation that these are entirely familiar to readers of this book, we illustrate only one of them with an example. In Figure 1.2, if we somehow know that $ST = YZ$ and that $\angle S = \angle Y$ and $\angle R = \angle X$, we can write in a proof: "We conclude by SAA that $\triangle SRT \cong \triangle YXZ$." (Notice that we wrote $\angle S$ as a short form for $\angle RST$, which is the full name of this angle. This is acceptable when it cannot result in ambiguity.)

What is the logical status of the four congruence criteria: SSS, SAS, ASA, and SAA? For each of these, the fact that the criterion is sufficient to guarantee the congruence of

GEOMETRY
for
College Students

The Brooks/Cole Series in Advanced Mathematics
Paul J. Sally, Jr., Editor

Probability: The Science of Uncertainty with Applications to Investments, Insurance, and Engineering
Michael A. Bean
University of Western Ontario
©2001 ISBN: 0534366031

The Mathematics of Finance: Modeling & Hedging
Joseph Stampfli
Victor Goodman
University of Indiana, Bloomington
©2001 ISBN: 0534377769

Geometry for College Students
I. Martin Isaacs
University of Wisconsin, Madison
©2001 ISBN: 0534351794

A Course in Approximation Theory
Ward Cheney
The University of Texas, Austin
Will Light
University of Leicester, England
©2000 ISBN: 0534362249

Introduction to Analysis, Fifth Edition
Edward D. Gaughan
New Mexico State University
©1998 ISBN: 0534351778

Numerical Analysis, Second Edition
David Kincaid
Ward Cheney
The University of Texas, Austin
©1996 ISBN: 0534338925

Advanced Calculus, A Course in Mathematical Analysis
Patrick M. Fitzpatrick
University of Maryland
©1996 ISBN: 0534926126

Algebra: A Graduate Course
I. Martin Isaacs
University of Wisconsin, Madison
©1994 ISBN: 0534190022

Fourier Analysis and Its Applications
Gerald B. Folland
University of Washington
©1992 ISBN: 0534170943

two triangles is actually a theorem, proved by Euclid from his postulates. These four theorems are among the basic results that we are accepting as known to be valid and that we are willing to use without providing proofs. In fact, however, it is not hard to prove the sufficiency of some of these criteria if we are willing to accept some of the others. As an example, and as the first proof that we actually present in this book, let us deduce the sufficiency of the SSS criterion, with the understanding that we may freely use any of the other three triangle-congruence conditions.

(1.1) **PROBLEM.** Assume that in $\triangle ABC$ and $\triangle RST$, we know that $AB = RS$, $AC = RT$, and $BC = ST$. Prove that $\triangle ABC \cong \triangle RST$ without using the SSS congruence criterion.

To do this, we shall appeal to another theorem that readers surely remember from their previous study of geometry: The base angles of an isosceles triangle are equal. Recall that a triangle $\triangle UVW$ is **isosceles** if two of its sides have equal lengths. In Figure 1.3, for example, the triangle is isosceles because $UV = UW$. The third side VW is called the **base** of the triangle, whether or not it actually occurs at the bottom of the diagram. The **base angles** of an isosceles triangle are the two angles at the ends of the base, and the theorem asserts that they are necessarily equal. In Figure 1.3, therefore, we have $\angle UVW = \angle UWV$. We mention that this base-angles theorem for isosceles triangles is used so often that it is given a name: the **pons asinorum**. This Latin phrase means "bridge of asses." Apparently, the theorem has acquired this name partly because the diagram in *The Elements* used in its proof vaguely resembles a bridge.

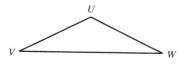

Figure 1.3

Solution to Problem 1.1. By renaming the points, if necessary, we can assume that AC is the longest side of $\triangle ABC$, and consequently, RT is the longest side of $\triangle RST$. Since we are given that $AC = RT$, we can move $\triangle RST$, flipping it over, if necessary, so that points A and R coincide, as do points C and T, and so that points B and S lie on opposite sides of line AC.

Now draw line segment BS. What must result is a situation resembling the left diagram in Figure 1.4. It is not possible for BS to fail to meet AC, as in the right diagram of Figure 1.4, or for BS to go through one of the points A or C because that would require one of BC or BA to be longer than AC. (We are shamelessly relying on the diagram to see this.) We can thus assume that we are in the situation of the left figure.

Since $AB = RS$, we see that $\triangle ABS$ is isosceles with base BS, and hence by the pons asinorum, we deduce that $x = y$, where, as indicated in the diagram, we are writing x and y to represent the measures of $\angle ABS$ and $\angle RSB$, respectively. Similarly, by a second application of the pons asinorum, we obtain $u = v$, and it

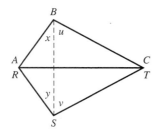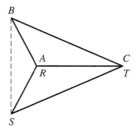

Figure 1.4

follows by addition that $x + u = y + v$. In other words, $\angle ABC = \angle RST$. Since we already know that $AB = RS$ and $BC = ST$, we can conclude by SAS that $\triangle ABC \cong \triangle RST$, as required.   ∎

Observe that we have written this proof in a conversational style, using complete English sentences grouped into logical paragraphs. We ended by establishing what we set out to prove, and we clearly marked the end of the proof. (It has become customary for a box to replace the old-fashioned QED as an end-of-proof marker.) This style, with minor variations, has become the accepted model for what a proof should look like throughout most areas of mathematics, and we follow it consistently in this book. We expect students to do the homework exercises with proofs written in the same style, using complete sentences. We do not recommend the two-column proof format that is often required in high-school geometry classes.

Note that the second sentence of the last paragraph of the proof begins with the word *similarly*. This word can be a powerful tool for simplifying and shortening proofs and making them more intelligible. Like most powerful tools, however, this one can be dangerous if used improperly. We encourage students to use the word *similarly* to avoid unnecessary repetition in their proofs, but to use it carefully.

Although we expect that most readers of this book remember how to prove the pons asinorum, we present more than one proof here to illustrate a few points and to provide further models of proof-writing style. Our first proof also yields some additional information about isosceles triangles. We remind the reader that a **median** of a triangle is the line segment joining a vertex to the midpoint of the opposite side.

(1.2) THEOREM.   *Let $\triangle ABC$ be isosceles, with base BC. Then $\angle B = \angle C$. Also, the median from vertex A, the bisector of $\angle A$, and the altitude from vertex A are all the same line.*

Proof.   In Figure 1.5, we have drawn the bisector AX of $\angle A$, and thus $\angle BAX = \angle CAX$. By hypothesis, we know that $AB = AC$, and of course, $AX = AX$. Thus $\triangle BAX \cong \triangle CAX$ by SAS. It follows that $\angle B = \angle C$ since these are corresponding angles in the congruent triangles.

We also need to show that the angle bisector AX is a median and that it is an altitude too. To see that it is a median, it suffices to check that X is the midpoint

of segment BC, and this is true because $BX = XC$ since these are corresponding sides of our congruent triangles. Finally, to prove that AX is also an altitude, we must show that AX is perpendicular to BC. In other words, we need to establish that $\angle BXA = 90°$. From the congruent triangles, we know that the corresponding angles $\angle BXA$ and $\angle CXA$ are equal, and thus $\angle BXA = \frac{1}{2}\angle BXC = 90°$ since the straight angle $\angle BXC = 180°$.   ∎

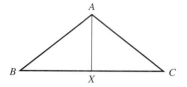

Figure 1.5

To prove that the angle bisector, median, and altitude from vertex A are all the same, we started by drawing one of these lines (the bisector) and showing that it was also a median and an altitude. Since there is only one median and one altitude from A, we know that the bisector is *the* median and *the* altitude.

What would have happened if we had started by drawing the median from A instead of the bisector of $\angle A$? This is, after all, the same line. In this situation, we could deduce that $\triangle BAX \cong \triangle CAX$ by SSS. It would then follow that $\angle BAX = \angle CAX$. We would deduce that AX is the angle bisector, and everything would proceed as before. This approach would have been less satisfactory, because it makes the pons asinorum depend on the SSS congruence criterion. In Problem 1.1, however, we proved the validity of the SSS criterion using the pons asinorum, and this would be an example of invalid circular reasoning.

What if we had started by drawing *altitude AX*? We would then know that $\angle BXA = \angle CXA$ since both of these are right angles. We also know that $AB = AC$ and $AX = AX$. At this point, we might be tempted to conclude that $\triangle BAX \cong \triangle CAX$ by SSA, but we would resist that temptation, of course, because SSA is not a valid congruence criterion. To see why, consider Figure 1.6. In this figure, the base BC of isosceles $\triangle ABC$ has been extended to an arbitrary point D beyond C. The two triangles $\triangle ADC$

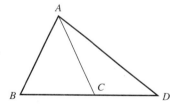

Figure 1.6

and $\triangle ADB$ are clearly not congruent because $DB > DC$, and yet the triangles agree in side-side-angle since $AB = AC$, $AD = AD$, and $\angle D = \angle D$.

There is one case where SSA is a valid congruence criterion: when the angle is a right angle. This is the "hypotenuse-arm" criterion, abbreviated HA. Recall that the longest side of a right triangle, the side opposite the right angle, is called the **hypotenuse** of the triangle. The other two sides of the triangle are often called its **arms**.

(1.3) THEOREM.   *If two right triangles have equal hypotenuses and an arm of one of the triangles equals an arm of the other, then the triangles are congruent.*

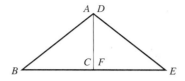

Figure 1.7

Proof.   We are given triangles $\triangle ABC$ and $\triangle DEF$ with right angles at C and F. We know that $AB = DE$ and $AC = DF$, and we want to show that $\triangle ABC \cong \triangle DEF$. Move $\triangle DEF$, flipping it over if necessary, so that points A and D and points C and F coincide and the diagram resembles Figure 1.7.

In Figure 1.7, we have $\angle BCE = \angle BCA + \angle EFD = 90° + 90° = 180°$, and thus BCE is a line segment, which we can now call BE. Since $AB = DE$, we see that $\triangle ABE$ is isosceles with base BE. Thus altitude AC is a median by Theorem 1.2, and hence $BC = FE$. The desired congruence now follows by SSS.   ■

We end this section with yet another proof of the pons asinorum. This one is amusing and very short, but it is somewhat tricky and should be read carefully.

Proof of pons asinorum.   We are given isosceles $\triangle ABC$ with base BC, and we want to show that $\angle B = \angle C$. We have $AB = AC$ and $AC = AB$. Since also $\angle A = \angle A$, we can conclude that $\triangle ABC \cong \triangle ACB$ by SAS. It follows that $\angle B = \angle C$ since these are corresponding angles of the congruent triangles.   ■

Exercises 1B

1B.1 Prove the converse of the pons asinorum. Show, in other words, that if $\angle B = \angle C$ in $\triangle ABC$, then $AB = AC$.

1B.2 If the altitude from vertex A in $\triangle ABC$ is also the bisector of $\angle A$, show that $AB = AC$.

1B.3 If the altitude from vertex A in $\triangle ABC$ is also a median, show that $AB = AC$.

   NOTE: This fact will be used later, so please do this problem.

1B.4 In Figure 1.8, medians BY and CZ have been drawn in isosceles $\triangle ABC$ with base BC. Show that $BY = CZ$.

1B.5 Using Figure 1.8 again, assume now that BY and CZ are angle bisectors of isosceles $\triangle ABC$ with base BC. Show that $BY = CZ$.

1B.6 Again in Figure 1.8, assume that $\triangle ABC$ is isosceles with base BC, but this time, assume that BY and CZ are altitudes. Show that $BY = CZ$.

   NOTE: Assume for this problem that, as in the figure, the two altitudes actually lie inside the triangle but observe that this does not always happen.

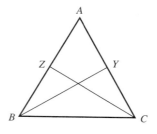

Figure 1.8

1B.7 Now in Figure 1.8, assume that BY and CZ are equal altitudes. Show that $AB = AC$.

1B.8 Referring again to Figure 1.8, let P be the point where BY and CZ meet. Assume that $BY = CZ$ and $PY = PZ$. Show that $AB = AC$.

1B.9 Once more in Figure 1.8, let P be the intersection point of BY and CZ. Assume that $AB = AC$ and $BY = CZ$. We ask if it is necessarily true that $PY = PZ$. You are asked, in other words, either to prove that $PY = PZ$ or else to show how to draw a counterexample diagram where all of the hypotheses hold but where PY and PZ clearly have different lengths. We drew Figure 1.8 so that PY and PZ actually are equal, and so if a counterexample exists, it would necessarily have to look somewhat different from the figure.

1C   Angles and Parallel Lines

A **transversal** is a line that cuts across two given lines. Usually, the two given lines are parallel when we use the word *transversal*, but we do not absolutely insist on this. In Figure 1.9, for example, line t is a transversal to lines a and b.

There is some nomenclature that is useful for discussing the eight angles that we have labeled with 1 through 8 in Figure 1.9. Angles that are on the same side of the transversal and on corresponding sides of the two lines a and b are said to be **corresponding** angles. In Figure 1.9, therefore, $\angle 1$ and $\angle 5$ are corresponding angles, as are $\angle 2$ and $\angle 6$ and also $\angle 3$ and $\angle 7$ and, of course, $\angle 4$ and $\angle 8$. It is a theorem that corresponding angles are equal when a transversal cuts a pair of parallel lines, and conversely, if in Figure 1.9, any one

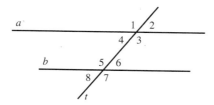

Figure 1.9

of the equalities $\angle 1 = \angle 5$, $\angle 2 = \angle 6$, $\angle 3 = \angle 7$, or $\angle 4 = \angle 8$ is known to hold, then it is a theorem that lines a and b are parallel, and thus the other three equalities also hold.

Pairs of angles such as $\angle 4$ and $\angle 6$ or $\angle 3$ and $\angle 5$ that lie on opposite sides of the transversal and between the two given lines are called **alternate interior** angles, and pairs such as $\angle 1$ and $\angle 7$ or $\angle 2$ and $\angle 8$ that lie on opposite sides of the transversal and outside of the space between the two parallel lines are **alternate exterior** angles. It is a theorem that alternate interior angles are equal and that alternate exterior angles are equal when a transversal cuts two parallel lines. It is also true that, conversely, if in Figure 1.9 any one of the equalities $\angle 1 = \angle 7$, $\angle 2 = \angle 8$, $\angle 3 = \angle 5$, or $\angle 4 = \angle 6$ is known to hold, then lines a and b must be parallel, and thus the other three equalities also hold, as do the four equalities mentioned in the previous paragraph.

We recall also that when two lines cross, as do lines a and t in Figure 1.9, then $\angle 1$ and $\angle 3$ are said to be **vertical angles**, as are $\angle 2$ and $\angle 4$. Vertical angles, of course, are always equal. While reviewing nomenclature for angles, we mention that two angles whose measures sum to $180°$ are said to be **supplementary**, and if the sum is $90°$, the angles are **complementary**. Of course, an angle of $180°$ is a **straight angle**, and an angle of $90°$ is a **right angle**. In Figure 1.9, we see that $\angle 1$ and $\angle 4$ are supplementary, and so if a and b are parallel, then $\angle 1 = \angle 5$, and it follows that $\angle 4$ and $\angle 5$ are supplementary, as are $\angle 3$ and $\angle 6$.

We can apply some of this to the angles of a triangle. Given $\triangle ABC$, extend side BC to point D, as shown in Figure 1.10. In this situation, $\angle ACD$ is said to be an **exterior** angle of the triangle at vertex C, and the two angles $\angle A$ and $\angle B$ are the **remote** interior angles with respect to this exterior angle.

(1.4) THEOREM.   *An exterior angle of a triangle equals the sum of the two remote interior angles. Also, the sum of all three interior angles of a triangle is* $180°$.

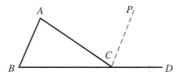

Figure 1.10

Proof.   We need to show in Figure 1.10 that $\angle ACD = \angle A + \angle B$. Draw a line CP through C and parallel to AB, as shown. Now $\angle A = \angle ACP$ since these are alternate interior angles for parallel lines AB and PC with respect to the transversal AC. Also, $\angle B = \angle PCD$ since these are corresponding angles. It follows that $\angle ACD = \angle ACP + \angle PCD = \angle A + \angle B$, as required.

We see in Figure 1.10 that $\angle ACD + \angle ACB = \angle BCD = 180°$. The substitution of $\angle A + \angle B$ for $\angle ACD$ in this equation yields the desired conclusion that the sum of the three interior angles of $\triangle ABC$ is $180°$.   ∎

Observe that our proof that the interior angles of a triangle total a straight angle relies on the fact that it is possible to draw a line through C parallel to AB. In fact,

Contents

CHAPTER 4

Ceva's Theorem and Its Relatives   *125*

CHAPTER 5

Vector Methods of Proof   *156*

CHAPTER 6

Geometric Constructions   *182*

GEOMETRY

for

College Students

there does not exist a proof of this result that does not, somehow, depend on parallel lines. This is because in non-Euclidean geometries, it is not true that the angles of a triangle must total 180°, and yet the only fundamental difference between Euclidean and non-Euclidean geometries is in the parallel postulate.

A triangle, of course, is a polygon with three sides. We digress to consider the question of how to find the sum of the interior angles of a polygon with n sides, where n may be larger than three.

(1.5) PROBLEM.   Find a formula for the sum of the interior angles of an n-gon.

Consider, for example, the case $n = 6$. In Figure 1.11, we see two 6-gons, which are usually called hexagons. In the left hexagon, we have drawn the three diagonals from vertex A. (A **diagonal** of a polygon is a line segment joining two of its nonadjacent vertices.) In general, an n-gon has exactly $n - 3$ diagonals terminating at each of its n vertices, and this gives a total of $n(n - 3)/2$ diagonals in all. (Do you see why we had to divide by 2 here?) A polygon is **convex** if all of its diagonals lie entirely in the interior. The interior angles of a polygon are the angles as seen from inside, and for a convex polygon such as the left hexagon in Figure 1.11, these angles are all less than 180°.

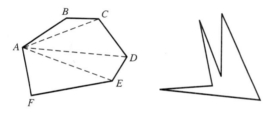

Figure 1.11

In the right hexagon of Figure 1.11, which is not convex, we see that two of its six interior angles exceed 180°. (An angle with measure larger than 180° is said to be a **reflex** angle.) In general, a polygon is convex if and only if its interior angles all measure less than 180°. If some interior angle of an n-gon is exactly equal to 180°, we do not consider it to be a convex n-gon, but in this situation, there are two adjacent sides that together form a single line segment, and the n-gon can be considered to be an $(n - 1)$-gon. Viewed as an $(n - 1)$-gon, the given polygon may be convex.

Suppose we have a convex n-gon such as hexagon $ABCDEF$ of Figure 1.11. Fix some particular vertex A and draw the $n - 3$ diagonals from A. This divides the original polygon into exactly $n - 2$ triangles, and it should be clear that the sum of all the interior angles of all of these triangles is exactly the sum of all interior angles of the original polygon. It follows that the sum of the interior angles of a convex polygon is exactly $180(n - 2)$ degrees.

We have not yet fully solved Problem 1.5, of course, because we have only considered convex polygons. Actually, it is not quite necessary for the polygon to be convex to make the previous argument work. What is really required is that there should exist at least one vertex from which all of the diagonals lie inside the polygon. Recall that the definition of a convex polygon requires that all diagonals from all vertices should be interior. It is conceivable that every polygon has at least one vertex from which all diagonals are interior, but unfortunately, that is not true. A careful inspection of the right hexagon in Figure 1.11 shows that it is a counterexample; at least one diagonal from every one of its vertices fails to be interior. It is true, but not easy to prove, that every polygon has at least one interior diagonal, and it is possible to use this hard theorem to construct a proof that for every n-gon, the sum of the interior angles is $180(n - 2)$ degrees.

There is another way to think about Problem 1.5 that may give additional insight. Imagine walking clockwise around a convex polygon, starting from some point other than a vertex on one of the sides. Each time you reach a vertex, you must turn right by a certain number of degrees. If the interior angle at the kth vertex is θ_k, then it is easy to see that the right turn at that vertex is a turn through precisely $180 - \theta_k$ degrees. When you return to your starting point, you will be facing in the same direction as when you started, and it should be clear that you have turned clockwise through a total of exactly $360°$. In other words,

$$\sum_{k=1}^{n}(180 - \theta_k) = 360.$$

Since the quantity 180 is added n times in this sum and each quantity θ_k is subtracted once, we see that $180n - \sum \theta_k = 360$, and hence $\sum \theta_k = 180n - 360 = 180(n-2)$. This provides a second proof of the formula for the sum of the interior angles of a convex n-gon.

How important is convexity for this second argument? If when walking clockwise around the polygon you reach the kth vertex and see a reflex interior angle there, you actually turn left, and not right. In this case, your left turn is easily seen to be through exactly $\theta_k - 180$ degrees, where, as before, θ_k is the interior angle at the vertex. If we view a left turn as being a right turn through some negative number of degrees, we see that at the kth vertex we are turning right by $180 - \theta_k$ degrees, and this is true regardless of whether $\theta_k < 180$ as in the convex case or $\theta_k > 180$ at a reflex-angle vertex. It is also clearly true at a straight-angle vertex, where $\theta_k = 180$. The previous calculation thus works in all cases, and it shows that $180(n - 2)$ is the sum of the interior angles for every polygon, convex or not.

1D   Parallelograms

Among polygons, perhaps parallelograms are second in importance after triangles. Recall that, by definition, a **parallelogram** is a quadrilateral $ABCD$ for which $AB \| CD$ and $AD \| BC$. In other words, the opposite sides of the quadrilateral are parallel. It is also true that the opposite sides of a parallelogram are equal, but this is a consequence of the assumption that the opposite sides are parallel; it is not part of the definition.

(1.6) THEOREM.   *Opposite sides of a parallelogram are equal.*

Proof.   In Figure 1.12, we are given that $AB \| CD$ and $AD \| BC$, and our task is to show that $AB = CD$ and $AD = BC$. Draw diagonal BD and note that $\angle ABD = \angle CDB$ since these are alternate interior angles for the parallel lines AB and CD, and similarly, $\angle DBC = \angle ADB$. Since $BD = BD$, we see that $\triangle DAB \cong \triangle BCD$ by ASA, and it follows that $AB = CD$ and $AD = BC$, as required. ■

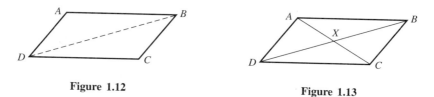

Figure 1.12   **Figure 1.13**

There are two useful converses for Theorem 1.6.

(1.7) THEOREM.   *In quadrilateral $ABCD$, suppose that $AB = CD$ and $AD = BC$. Then $ABCD$ is a parallelogram.*

(1.8) THEOREM.   *In quadrilateral $ABCD$, suppose that $AB = CD$ and $AB \| CD$. Then $ABCD$ is a parallelogram.*

We leave the proofs of these two theorems to the exercises.

(1.9) THEOREM.   *A quadrilateral is a parallelogram if and only if its diagonals bisect each other.*

Proof.   As shown in Figure 1.13, we let X be the point where diagonals AC and BD of quadrilateral $ABCD$ cross. Suppose first that X is the common midpoint of line segments AC and BD. Then $AX = XC$ and $BX = XD$ and also $\angle AXB = \angle CXD$ because these are vertical angles. It follows that $\triangle AXB \cong \triangle CXD$ by SAS, and thus $AB = CD$. Similarly, $AD = BC$, and hence $ABCD$ is a parallelogram by Theorem 1.7.

Conversely now, we assume that $ABCD$ is a parallelogram, and we show that X is the midpoint of each of the diagonals AC and BD. We have $\angle BAX = \angle XCD$ and $\angle ABX = \angle CDX$ because in each case, these are pairs of alternate interior angles for the parallel lines AB and CD. Also, $AB = CD$ by Theorem 1.6, and thus $\triangle ABX \cong \triangle CDX$ by ASA. We deduce that $AX = CX$ and $BX = DX$, as required. ■

Observe that by Theorem 1.7, a quadrilateral in which all four sides are equal must be a parallelogram. Such a figure is called a **rhombus**. In the case of a rhombus, the diagonals not only bisect each other, but they are also perpendicular. Although this fact is not difficult to prove directly, we prefer to derive it as a consequence of the following more general result.

(1.10) THEOREM. *Given a line segment BC, the locus of all points equidistant from B and C is the perpendicular bisector of the segment.*

Proof. We need to show that every point on the perpendicular bisector of BC is equidistant from B and C, and we must also show that every point that is equidistant from B and C lies on the perpendicular bisector of BC.

Assume that AX is the perpendicular bisector of BC in Figure 1.14. This means that X is the midpoint of BC and AX is perpendicular to BC. In other words, we are assuming that AX is simultaneously a median and an altitude in $\triangle ABC$, and we want to deduce that $AB = AC$. This is precisely Exercise 1B.3, and we will not give the proof here.

Assume now that A is equidistant from B and C in Figure 1.14 and draw median AX of $\triangle ABC$. Since $AB = AC$, this triangle is isosceles, and thus by Theorem 1.2, median AX is also an altitude. In other words, AX is the perpendicular bisector of BC, and of course, A lies on this line. ∎

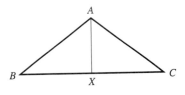

Figure 1.14

(1.11) COROLLARY. *The diagonals of rhombus $ABCD$ are perpendicular.*

Proof. Since $AB = AD$, we know by Theorem 1.10 that A lies on the perpendicular bisector of diagonal BD, and similarly, C lies on the perpendicular bisector of BD. But AC is the only line that contains the two points A and C, and thus AC is the perpendicular bisector of BD. In particular, diagonal AC is perpendicular to diagonal BD. ∎

We close this section by mentioning another special type of parallelogram: a rectangle. By definition, a **rectangle** is a quadrilateral all of whose angles are right angles. It is easy it see that the opposite sides of a rectangle are parallel, and so a rectangle is automatically a parallelogram. We also know (by Problem 1.5, for example) that the sum of the four angles of an arbitrary quadrilateral is 360°. Thus if we have any quadrilateral with all four angles equal, each angle must be 90°, and the figure is a rectangle. We remark also that adjacent vertices of a parallelogram have supplementary interior angles. It follows easily that if one angle of a parallelogram is a right angle, then the parallelogram must be a rectangle. Finally, we mention that a quadrilateral that is both a rectangle and a rhombus is, by definition, a **square**.

Exercises 1D

1D.1 Prove Theorem 1.7 by showing that quadrilateral $ABCD$ is a parallelogram if $AB = CD$ and $AD = BC$.

1D.2 Prove Theorem 1.8 by showing that quadrilateral $ABCD$ is a parallelogram if $AB = CD$ and $AB \| CD$.

1D.3 Prove that the diagonals of a rectangle are equal.

1D.4 Prove that a parallelogram having perpendicular diagonals is a rhombus.

1D.5 Prove that a parallelogram with equal diagonals is a rectangle.

1D.6 In Figure 1.15, we are given that $AB \| CD$ and that AD and BC are not parallel. Show that $\angle D = \angle C$ if and only if $AD = BC$.

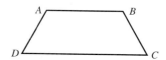

Figure 1.15

HINT: Draw a line through B parallel to AD.

NOTE: Recall that a quadrilateral is said to be a **trapezoid** if it has exactly one pair of parallel sides. If the nonparallel pair of sides are equal, the trapezoid is said to be **isosceles**.

1D.7 Show that opposite angles of a parallelogram are equal.

1D.8 In quadrilateral $ABCD$, suppose that $AB \| CD$ and $\angle B = \angle D$. Show that $ABCD$ is a parallelogram.

1D.9 In quadrilateral $ABCD$, suppose that $\angle A = \angle C$ and $\angle B = \angle D$. (We are referring to the interior angles, of course.) Show that $ABCD$ is a parallelogram.

HINT: Show that $\angle A$ and $\angle C$ are supplementary.

1D.10 Show that the diagonals of an isosceles trapezoid are equal.

1D.11 In Figure 1.16, we are given that $\triangle AXB$ and $\triangle AYB$ are isosceles and share base AB. Show that points X, Y, and Z are collinear (lie on a common line) if and only if $\triangle AZB$ is isosceles with base AB.

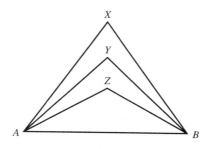

Figure 1.16

1D.12 In Figure 1.17, point O is equidistant from the vertices of $\triangle ABC$. This point is reflected in each side of the triangle, yielding points X, Y, and Z. This means that OX is perpendicular to BC and that the perpendicular distance from X to BC is equal to that from O to BC. Similar assertions hold for Y and Z. Prove that $\triangle ABC \cong \triangle XYZ$ and that corresponding sides of these two triangles are parallel.

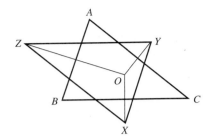

Figure 1.17

HINT: Show that quadrilaterals $ZBOA$ and $COAY$ are rhombuses. Deduce that BZ and CY are equal and parallel.

NOTE: Given any triangle, there always exists a point equidistant from the three vertices. This point, called the **circumcenter** of the triangle, does not always lie inside the triangle. The assertion of this problem is still valid if O lies outside of $\triangle ABC$ or even if it lies on one of the sides.

1D.13 Each pair of parallel lines in Figure 1.18 represents a railroad track with two parallel rails. The distances between the rails in each of the tracks are equal. Prove that the parallelogram formed where the tracks cross is a rhombus.

Figure 1.18

NOTE: The distance between the rails is, of course, measured perpendicularly. It should be clear that for each track, the perpendicular distance from a point on one rail to the other rail is constant, independent of the point.

1D.14 Recall that the distance from a point to a line is measured perpendicularly. Given $\angle ABC$, show that the locus of all points inside the angle and equidistant from the two lines BA and BC is the bisector of the angle.

1E   Area

We assume as known the fact that the area of a rectangle is the product of its length and width. The area of a geometric figure is often denoted K, and so the formula for the area of a rectangle is $K = bh$, where b is the length of one of the sides of the rectangle and h is the length of the perpendicular sides. A side of length b is referred to as the base of the rectangle, and the height h is the length of the sides perpendicular to the base. Of course, the base of a rectangle does not have to be at the bottom; it can be any side.

Now suppose that we have a parallelogram that is not necessarily a rectangle. Again we designate one side of the parallelogram as the base and write b to denote its length. But this time, the height h is the perpendicular distance between the two parallel

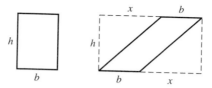

Figure 1.19

sides of length b. Of course, if the parallelogram happens to be a rectangle, then this perpendicular distance is the length of the other two sides.

On the left of Figure 1.19, we see a rectangle with base b and height h. On the right, we have drawn a parallelogram that also has base b and height h, and we claim that the parallelogram and the rectangle have equal areas. To see why this is true with a very informal argument, drop perpendiculars from one end of each of the sides of length b of the parallelogram to the extensions of the opposite sides, as shown in the figure. What results is a rectangle with base $b + x$ and height h, where, as shown, x represents the amount that the base of the parallelogram had to be extended to meet the perpendicular. The area of this rectangle is $(b + x)h$, and to obtain the area of the original parallelogram, we need to subtract from this the area of the two right triangles. Clearly, the two right triangles pasted together would form a rectangle with base x and height h, and so the total area of the two triangles is xh. The area of the parallelogram is therefore $(b + x)h - xh = bh$, and so the formula $K = bh$ works to find areas of arbitrary parallelograms and not just of rectangles.

Another way to see why the rectangle and parallelogram in Figure 1.19 have equal area is from the point of view of calculus. Imagine slicing each of the areas into infinitesimally thin horizontal strips of length b. We see that each of the two areas is given by the same formula

$$K = \int_0^h b \, dy,$$

and thus the two areas are equal, as claimed. In fact, if we carry out the integration, we get the formula $K = bh$, as we expect.

Next, we consider areas of triangles. In the left diagram of Figure 1.20, we have drawn a triangle with base b and height h, where any one of the three sides can be viewed as the base and the length of the altitude drawn to that side is the corresponding height. Remember that this altitude may lie outside of the triangle.

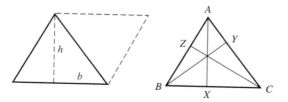

Figure 1.20

To compute the area K of the given triangle, we have constructed a parallelogram by drawing lines parallel to our base and to one of the other sides of the triangle, as shown in the figure. This parallelogram has base b and height h, and so its area is bh. The parallelogram is divided into two congruent triangles by a diagonal, and so the area of each of these triangles is exactly half the area of the parallelogram. (The two triangles are easily seen to be congruent by SSS.) Since the original triangle is one of the two halves of the parallelogram, we see that $K = \frac{1}{2}bh$, and this is the basic formula for the area of a triangle.

As is indicated with $\triangle ABC$ on the right of Figure 1.20, any one of the three sides could have been designated as the base, and for each, there is a corresponding altitude. (It is not a coincidence that the three altitudes are concurrent; we shall see in Chapter 2 that this is guaranteed to happen.) We thus have three different formulas for the the area K_{ABC} of $\triangle ABC$. We have

$$K_{ABC} = \frac{AX \cdot BC}{2} = \frac{BY \cdot AC}{2} = \frac{CZ \cdot AB}{2} \, ,$$

and we can deduce useful information relating the lengths of the sides and altitudes of a triangle. For example, in Exercise 1B.6, you were asked to show that in an isosceles $\triangle ABC$ with base BC, the two altitudes BY and CZ are equal, and you were allowed to assume that the altitudes were inside the triangle. By the area formula we have just derived, we know that $BY \cdot AC = CZ \cdot AB$ since each of these quantities equals twice the area of the triangle. We can cancel the equal quantities AC and AB to obtain $BY = CZ$, as desired. This proof, furthermore, works independently of whether or not the altitudes lie inside the triangle. We remark also that the converse is true: If we are given that altitudes BY and CZ are equal, it follows from the formula $BY \cdot AC = CZ \cdot AB$ that $AB = AC$. (This converse appeared as Exercise 1B.7.)

We can get some additional information about areas of triangles by using a little elementary trigonometry. In Figure 1.21, we have followed custom and used the symbols a, b, and c to denote the lengths of the sides of $\triangle ABC$ opposite vertices A, B, and C, respectively. Also, we have drawn the altitude of length h from A. (In the figure, this altitude happens to lie outside the triangle, but this is irrelevant to our calculation.) We have, of course, $K = \frac{1}{2}ah$, where, as usual, $K = K_{ABC}$ represents the area of the triangle. We see that $\sin(C) = h/b$, and hence $h = b\sin(C)$, and if we substitute this into the area formula $K = \frac{1}{2}ah$, we obtain $K = \frac{1}{2}ab\sin(C)$. Similarly, we also have $K = \frac{1}{2}ac\sin(B)$ and $K = \frac{1}{2}bc\sin(A)$.

It is now clear that for any triangle, the equations

$$ab\sin(C) = bc\sin(A) = ca\sin(B)$$

must always hold. If we divide through by the quantity abc and take reciprocals, we obtain the so-called law of sines:

$$\frac{c}{\sin(C)} = \frac{a}{\sin(A)} = \frac{b}{\sin(B)} \, .$$

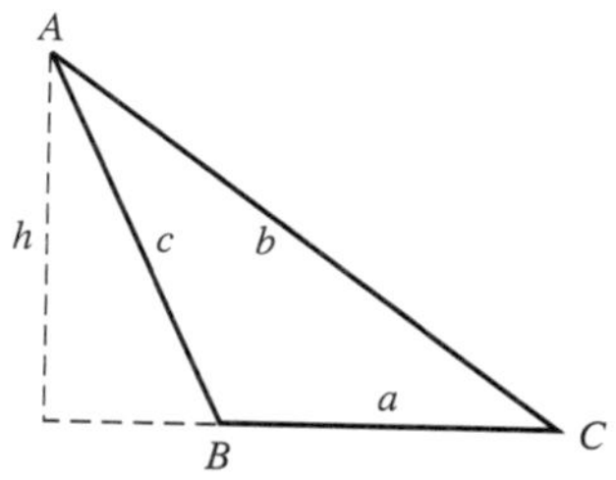

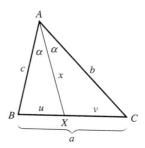

Figure 1.21

Figure 1.22

There is a simple proof of a more powerful formula called the extended law of sines that we prove in Chapter 2. Another application of the sine formula for the area of a triangle is the following.

(1.12) THEOREM.   *Let AX be the bisector of $\angle A$ in $\triangle ABC$. Then*

$$\frac{BX}{XC} = \frac{AB}{AC}.$$

In other words, X divides BC into pieces proportional to the lengths of the nearer sides of the triangle.

Proof.   In the notation of Figure 1.22, we must show that $u/v = c/b$, where b and c are as usual and u and v are the lengths of BX and XC, as shown. To see why this equation holds, let h be the height of $\triangle ABC$ with respect to the base BC. Then h is also the height of each of $\triangle ABX$ and $\triangle ACX$ with respect to bases BX and XC, respectively. We have

$$\frac{uh}{2} = K_{ABX} = \frac{cx \sin(\alpha)}{2} \quad \text{and} \quad \frac{vh}{2} = K_{ACX} = \frac{bx \sin(\alpha)}{2},$$

where $x = AX$ and we have written $\alpha = \frac{1}{2}\angle A$. Division of the first of these equations by the second yields the desired proportion.   ∎

As an application, we have the following, which should be compared with Exercises 1B.2 and 1B.3. Unlike those exercises, it it difficult to see a direct and elementary proof via congruent triangles for this fact.

(1.13) PROBLEM.   Suppose that in $\triangle ABC$, the median from vertex A and the bisector of $\angle A$ are the same line. Show that $AB = AC$.

Solution.   In the notation of Figure 1.22, we have $u = v$ since the angle bisector AX is assumed to be a median. It follows from Theorem 1.12 that $a = b$, as required.   ∎

We have now given two different types of formula for the area of a triangle: one using one side and an altitude and the other using two sides and an angle. Since the SSS congruence criterion tells us that a triangle is determined by its three sides, we might expect that there should be a nice way to compute the area of a triangle in terms of the lengths of its sides. There is. In Chapter 2, we prove the following formula, which is attributed to Heron of Alexandria (c. 50 A.D.):

$$K_{ABC} = \sqrt{s(s-a)(s-b)(s-c)}\,,$$

where $s = \frac{1}{2}(a+b+c)/2$ is called the **semiperimeter** of the triangle. Of course, a, b, and c retain their usual meanings.

We close this section with what seems to be an amazing fact.

(1.14) PROBLEM. As shown in Figure 1.23, points P, Q, and R lie on the sides of $\triangle ABC$. Point P lies one third of the way from B to C, point Q lies one third of the way from C to A, and point R lies one third of the way from A to B. Line segments AP, BQ, and CR subdivide the interior of the triangle into three quadrilaterals and four triangles, as shown, and we see that exactly one of the four small triangles has no vertex in common with $\triangle ABC$. Prove that the area of this triangle is exactly one seventh of the area of the original triangle.

Solution. We need to compute the area K_{XYZ} in Figure 1.23. By choosing units appropriately, we can assume that $K_{ABC} = 3$ and we write k to denote K_{BYP}. We draw line segment YC and start computing areas.

Since $\triangle CYP$ has the same height as $\triangle BYP$ but its base PC is twice as long, we deduce that $K_{CYP} = 2k$. Similarly, since $AQ = 2QC$, we see that $K_{ABQ} = 2K_{CBQ}$, and thus $K_{CBQ} = \frac{1}{3}K_{ABC} = 1$. This enables us to compute that $K_{CYQ} = 1 - K_{BYC} = 1 - 3k$. By the usual reasoning, $K_{AYQ} = 2K_{CYQ} = 2 - 6k$, and we know that $K_{ABQ} = \frac{2}{3}K_{ABC} = 2$. It follows that $K_{ABY} = 2 - K_{AYQ} = 2 - (2 - 6k) = 6k$. However, $K_{ABP} = \frac{1}{3}K_{ABC} = 1$, and thus $K_{BYP} = 1 - 6k$. But we know that $K_{BYP} = k$, and hence $1 - 6k = k$ and $k = 1/7$.

Similar reasoning shows that $K_{ARX} = 1/7 = K_{CQZ}$, and since $K_{ARC} = \frac{1}{3}K_{ABC} = 1$, we deduce that the area of quadrilateral $AXZQ$ is $1 - 2/7 = 5/7$. Finally, we recall that $K_{AYQ} = 2 - 6k = 8/7$, and it follows that $K_{XYZ} = 8/7 - 5/7 = 3/7$. This is exactly one seventh of the area of the original triangle, as desired. ∎

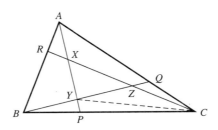

Figure 1.23

Exercises 1E

1E.1 Draw two of the medians of a triangle. This subdivides the interior of the triangle into four pieces: three triangles and a quadrilateral. Show that two of the three small triangles have equal area and that the area of the third is equal to that of the quadrilateral.

1E.2 An arbitrary point P is chosen on side BC of $\triangle ABC$ and perpendiculars PU and PV are drawn from P to the other two sides of the triangle. (It may be that U or V lies on an extension of AB or AC and not on the actual side of the triangle. This can happen, for instance, if $\angle A$ is obtuse and point P is very near B or C.) Show that the sum $PU + PV$ of the lengths of the two perpendiculars is constant as P moves along BC. In other words, this quantity is independent of the choice of P.

1E.3 Since a triangle is determined by angle-side-angle, there should be a formula for K_{ABC} expressed in terms of a and $\angle B$ and $\angle C$. Derive such a formula.

1E.4 Let P be an arbitrary point in the interior of a convex quadrilateral. Draw the line segments joining P to the midpoints of each of the four sides, thereby subdividing the interior into four quadrilaterals. Now choose two of the small quadrilaterals not having a side in common and show that the areas of these two total exactly half the area of the original figure.

1F   Circles and Arcs

As all readers of this book surely know, a **circle** is the locus of all points equidistant from some given point called the **center**. The common distance r from the center to the points of the circle is the **radius**, and the word *radius* is also used to denote any one of the line segments joining the center to a point of the circle. A **chord** is any line segment joining two points of a circle, and a **diameter** is a chord that goes through the center. The length d of any diameter is given by $d = 2r$, and this is the maximum of the lengths of all chords. Finally, we mention that any two circles with equal radii are congruent, and any point on one of two congruent circles can be made to correspond to any point on the other circle.

Just as any two points determine a unique line, it is also true that any three points, unless they happen to lie on a line, lie on a unique circle.

(1.15) THEOREM. *There is exactly one circle through any three given noncollinear points.*

Proof. Call the points A, B, and C. Since by hypothesis, there is no line through these points, we can be sure that we are dealing with three distinct points, and we draw line segments AB and AC. Let b and c be the perpendicular bisectors of these segments and observe that lines b and c cannot be parallel because lines AB and AC are neither parallel nor are they the same line. (Since AB and AC have point A in common, they surely are not parallel; they cannot be the same line because we are assuming that no line contains all three points A, B, and C.)

Let P be the point where lines b and c meet. Since P lies on the perpendicular bisector of AB, we know by Theorem 1.10 that P is equidistant from A and B. In other words, $PA = PB$. Similarly, since P lies on line c, we deduce that $PA = PC$. If we let r denote the common length of the three segments PA, PB, and PC, we see that the circle of radius r centered at P goes through the three given points.

To see that our three points cannot lie on any other circle, we could appeal to the "obvious" fact that two different circles meet in at most two points, but it is almost as easy to give a real proof. If a circle centered at some point Q, say, goes through A, B, and C, then Q is certainly equidistant from A and B, and hence by Theorem 1.10, it lies on the perpendicular bisector b of segment AB. Similarly, we see that Q lies on line c and thus $Q = P$ because P is the only point common to the two lines. Since the distance $PA = r$, it follows that the only circle through A, B, and C is the circle of radius r centered at P. ∎

Given $\triangle ABC$, the unique circle that goes through the three vertices is called the **circumcircle** of the triangle, and the triangle is said to be **inscribed** in the circle. More generally, any polygon all of whose vertices lie on some given circle is referred to as being **inscribed** in that circle, and the circle is **circumscribed** about the polygon. Although every triangle is inscribed in some circle, the same cannot be said for n-gons when $n > 3$.

The statement of Theorem 1.15 would be neater if we did not have to deal with the exceptional case where the three given points are collinear. If we were willing to say that a line is a certain kind of circle (which, of course, it is not), we could then say that every choice of three distinct points determines a circle. It is sometimes convenient to pretend that a line is a circle with "infinite" radius, but of course, this should not be taken too literally.

We return now to our study of genuine circles. Two points A and B on a circle divide the circle into two pieces, each of which is called an **arc**. We write $\overarc{AB}$ to denote one of these two arcs, usually the smaller. As this is ambiguous, a three-point designation for an arc is often preferable. In Figure 1.24, for example, the smaller of the two arcs determined by points A and B would be designated $\overarc{AXB}$ and the larger is $\overarc{AYB}$. The ambiguity in the notation $\overarc{AB}$ is related to a similar ambiguity in the notation for angles. For example, if we write $\angle AOB$, we generally mean the angle that in Figure 1.24 includes point X in its interior; we do not mean the reflex angle, with Y in its interior.

How big is an arc? The most common way to discuss the size of an arc is in terms of the fraction of the circle it is, where the whole circle is taken to be $360°$ or 2π radians. An arc extending over a quarter of the circle, therefore, is referred to as a $90°$ arc, and we would write $\overarc{AB} \overset{\circ}{=} 90°$ or $\overarc{AB} \overset{\circ}{=} \pi/2$ radians in this case. Of course, this size description for an arc is meaningful only relative to the circle of which it is a part. If we are told, for example, that we have two $90°$ arcs, we cannot say that they are congruent or that they have equal length unless we know that these are two arcs of the same circle or of two circles having equal radii. To remind us that the number of degrees (or radians) that we assign to an arc gives only relative information, we use the symbol $\overset{\circ}{=}$, which

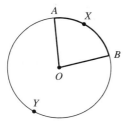

Figure 1.24

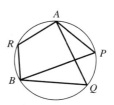

Figure 1.25

we read as "equal in degrees (or radians)," and we avoid the use of $=$ in this context. Informally, however, we do speak of equal arcs, but it is probably best to avoid this phrase except when we know that the two arcs are in the same or equal circles.

Given an arc $\overset{\frown}{AB}$ on a circle centered at point O, we say that $\angle AOB$ is the **central angle** corresponding to the arc. Since a full circle is 360° of arc and one full rotation is 360° of angle, it should be clear that the number of degrees in the measure of central angle $\angle AOB$ is equal to the number of degrees in $\overset{\frown}{AB}$. A 90° angle at the central point O, for example, cuts off a quarter circle, which is a 90° arc. The standard jargon for the phrase "cuts off" in the previous sentence is **subtends**. In general, we can write $\angle AOB \overset{\circ}{=} \overset{\frown}{AB}$. A central angle, therefore, is equal in degrees to the arc it subtends. We can also say that the arc is **measured** by the central angle.

In a given circle, the angle formed by two chords that share an endpoint is called an **inscribed** angle. Some of the inscribed angles in Figure 1.25, for example, are $\angle APB$, $\angle AQB$, and $\angle ARB$.

(1.16) THEOREM. *An inscribed angle in a circle is equal in degrees to one half its subtended arc. Equivalently, the arc subtended by an inscribed angle is measured by twice the angle.*

In Figure 1.25, for example, $\angle APB \overset{\circ}{=} \frac{1}{2} \overset{\frown}{ARB}$ and also $\angle AQB \overset{\circ}{=} \frac{1}{2} \overset{\frown}{ARB}$. In particular, we can deduce that $\angle APB = \angle AQB$, and in general, any two inscribed angles that subtend the same arc in a circle are equal. As we shall see, this provides a useful technique for proving equality of angles. Now consider $\angle ARB$ in Figure 1.25. Like $\angle APB$ and $\angle AQB$, this too is an inscribed angle formed by chords through A and B, but we cannot conclude that $\angle ARB$ is equal to the other two angles because it subtends the other arc determined by points A and B. In fact, $\angle ARB \overset{\circ}{=} \frac{1}{2} \overset{\frown}{APB}$. Since $\overset{\frown}{ARB}$ and $\overset{\frown}{APB}$ together constitute the whole circle, it follows that

$$\angle ARB + \angle APB = \frac{1}{2}(\overset{\frown}{APB} + \overset{\frown}{ARB}) = \frac{1}{2}360° = 180°,$$

and thus $\angle ARB$ and $\angle AQB$ are supplementary. This proves the following corollary to Theorem 1.16. Recall that a polygon is inscribed in a circle if all of its vertices lie on the circle.

(1.17) COROLLARY. *Opposite angles of an inscribed quadrilateral are supplementary.* ∎

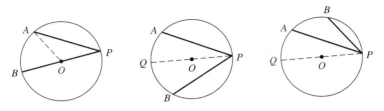

Figure 1.26

Proof of Theorem 1.16.   Given $\angle APB$ inscribed in a circle centered at point O, the three cases we need to consider are illustrated in Figure 1.26. It may be that point O falls on one of the sides of $\angle APB$, as in the left diagram. Alternatively, O might lie in the interior of the angle, as in the middle diagram of Figure 1.26, or it might be exterior to the angle, as in the right diagram.

Suppose first that O lies on a side (say, PB). Draw radius AO and observe that $\triangle AOP$ is isosceles with base AP, and so by the pons asinorum, $\angle A = \angle P$. Central $\angle AOB$ is an exterior angle of $\triangle AOP$, and hence it is equal to the sum of the two remote interior angles by Theorem 1.4. Thus

$$\widehat{AB} \overset{\circ}{=} \angle AOB = \angle A + \angle P = 2\angle P \,,$$

and hence $\angle P \overset{\circ}{=} \frac{1}{2} \widehat{AB}$, as required.

Now assume that the center O of the circle lies in the interior of the angle and draw diameter PQ, as in the middle diagram of Figure 1.26. By the part of the theorem that we have already proved, we know that

$$\angle APQ \overset{\circ}{=} \frac{1}{2} \widehat{AQ} \qquad \text{and} \qquad \angle QPB \overset{\circ}{=} \frac{1}{2} \widehat{QB} \,.$$

Adding these equalities gives $\angle APB \overset{\circ}{=} \frac{1}{2} \widehat{AQB}$, as required.

Finally, we can assume the situation of the right diagram in Figure 1.26. Again we draw diameter PQ and we get the same two equalities as in the previous case. This time, subtraction yields the desired result.   ∎

Imagine the following experiment. Mark a large circle on the ground and erect two vertical poles at points A and B on the circle. Now stand somewhere else on the circle and observe how far apart the two poles appear to be, as seen from your perspective. The apparent separation of the poles is determined by the angle from one pole to your eye and back to the other pole. Since this angle is inscribed in the circle, Theorem 1.16 guarantees that, as we walk along the circle, the apparent separation of the poles remains unchanged, as long as we stay on one of the two arcs determined by A and B. The angular separation we see from anywhere on the other arc determined by A and B, however, is the supplement of this angle.

Suppose now that the line through the two poles runs due north and south. We know that from all points on the circle and east of the poles, the separation between the poles appears constant. In Figure 1.27, in other words, $\angle APB$ is independent of the choice of the point P on the eastern arc of the circle.

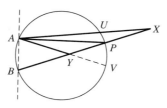

Figure 1.27

What happens if we remain east of the poles but move outside of the circle, to point X, for example? Common sense tells us that since X is farther from the poles than is P, the angular separation should decrease. We can quantify this because, as we see in the figure, $\angle APB$ is an exterior angle of $\triangle APX$. Thus $\angle AXB = \angle APB - \angle XAP$, and the apparent decrease in angular separation as we move from P to X is precisely equal to $\angle XAP$. Also, if we notice that $\angle XAP = \angle UAP \stackrel{\circ}{=} \frac{1}{2} \stackrel{\frown}{UP}$, we see that

$$\angle AXB \stackrel{\circ}{=} \frac{1}{2}(\stackrel{\frown}{AB} - \stackrel{\frown}{UP}).$$

A line segment (such as AX and BX) that extends a chord beyond a circle is called a **secant**, and so we have thus proved the following result.

(1.18) COROLLARY. *The angle between two secants drawn to a circle from an exterior point is equal in degrees to half the difference of the two subtended arcs.* ∎

Returning now to our two poles, we see in Figure 1.27 that if we move from point P to a point Y inside the circle and east of the poles, the apparent separation of the poles increases from $\angle APB$ to $\angle AYB$. Since $\angle AYB$ is an exterior angle of $\triangle AYP$, it follows that the amount of increase equals $\angle AYB - \angle APB = \angle PAY \stackrel{\circ}{=} \frac{1}{2} \stackrel{\frown}{PV}$. Thus

$$\angle AYB \stackrel{\circ}{=} \frac{1}{2}(\stackrel{\frown}{AB} + \stackrel{\frown}{PV}).$$

We have proved the following.

(1.19) COROLLARY. *The angle between two chords that intersect in the interior of a circle is equal in degrees to half the sum of the two subtended arcs.* ∎

We have seen that the part of the plane east of our two poles is subdivided into three sets. Everywhere on the circle, the angular separation of the poles is $\frac{1}{2} \stackrel{\frown}{AB}$. Inside the circle, the angular separation is always greater than this quantity, and outside the circle, the angular separation is always smaller. We close this discussion with a little exercise:

Given two points and some angle θ, sketch the locus of all points on the plane from which the angular separation of the two given points is equal to θ. Consider separately the three cases $\theta < 90°$, $\theta = 90°$, and $\theta > 90°$.

(1.20) PROBLEM.   A 6-foot-tall rectangular painting is hung high on a wall, with its bottom edge 7 feet above the floor. An art lover whose eyes are 5 feet above the floor wants as good a view as possible, and so she wants to maximize the angular separation from her eye to the top and the bottom of the painting. How far from the wall should she stand?

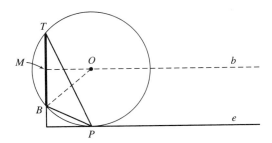

Figure 1.28

Solution.   In Figure 1.28, horizontal line e represents the possible positions of the viewer's eye, 5 feet above the floor. Line TB is the wall on which the picture is hung, and T and B represent, respectively, the top and bottom of the picture. We seek a point P on line e that maximizes $\angle BPT$. Although this can be done by calculus techniques, we present an easier method.

We know that point B is 2 feet above line e and that T is 6 feet higher, or 8 feet above e. The midpoint M of TB is thus at the average height $(2+8)/2 = 5$ feet above e. Draw the perpendicular bisector b of TB so that b is horizontal and 5 feet above e and choose point O on b so that $OB = 5$. Draw the circle of radius 5 centered at O and note that this circle is tangent to e at some point P. The circle, in other words, touches line e at P, but it does not extend below e.

We argue that this point P solves our problem. Every other point on line e lies outside the circle and thus "sees" the picture TB with a smaller angle than does P. In other words, $\angle BPT$ is the maximum we seek. To answer the question that was asked, we need to know how far point P is from the wall. This distance is equal to OM, and so we examine the right triangle $\triangle OMB$. We know that hypotenuse $OB = 5$ because OB is a radius of the circle. Also, $MB = 3$, and hence by the Pythagorean theorem, we see that $OM = 4$. The answer to the problem, therefore, is that the art lover should stand 4 feet from the wall.   ■

Since we used the Pythagorean theorem in the solution to Problem 1.20, perhaps this is a good place to digress to give an elegant noncomputational proof.

(1.21) THEOREM (Pythagoras). *If a right triangle has arms of lengths a and b and its hypotenuse has length c, then $a^2 + b^2 = c^2$.*

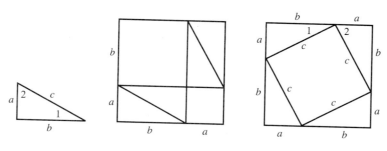

Figure 1.29

Proof. As we shall explain, the whole proof is visible in Figure 1.29. On the left, we see our given triangle. The middle diagram shows a square of side $a + b$ decomposed into two squares and four right triangles. We observe, furthermore, that each of the right triangles has arms of length a and b, and so each is congruent to the original triangle by SAS. The two smaller squares in the middle diagram have side lengths a and b, and so the area remaining in the big square of side $a + b$ when four copies of our given triangle are removed is exactly $a^2 + b^2$.

On the right, we see another square of side $a + b$ decomposed this time into one square and four right triangles. Again, each of the right triangles is congruent to the given one by SAS, and so the side length of the smaller square is exactly c. It follows that the area remaining when four copies of our triangle are removed from a square of side $a + b$ is c^2. We just saw, however, that this area is equal to $a^2 + b^2$, and thus $a^2 + b^2 = c^2$, as required.

There is one crucial detail we have omitted. It is clear that the "square" of side c in the figure on the right is a rhombus, but why are its angles equal to 90°? Look, for example, at the top of the diagram on the right. We see that the angle of the rhombus together with ∠1 and ∠2 make a straight angle of 180°. In the original triangle, however, we know that ∠1, ∠2, and a right angle sum to 180°, and it follows that the angle of the rhombus is 90°, as required. ∎

We mention an alternative argument that could be used to see why the rhombus with side length c in the right diagram of Figure 1.29 has to be a square. Observe that there is a rotational symmetry in the figure. In other words, if we rotate the entire large square by a quarter turn, what results is identical to the figure with which we started. It follows from this that all four angles of the rhombus must be equal. Since we know that the sum of the angles of any quadrilateral is 360°, we deduce that each corner of the rhombus is 90°, and thus the rhombus is a square.

We return now to our study of angles in circles. There is a special case of Theorem 1.16 that is used so often that it deserves special mention.

(1.22) COROLLARY. *Given △ABC, the angle at vertex C is a right angle if and only if side AB is a diameter of the circumcircle.*

Proof. We know that the circumcircle exists by Theorem 1.15. In this circle, $\overset{\frown}{AB}$ is measured by $2\angle C$. Here, of course, we have written $\overset{\frown}{AB}$ to denote the arc not containing C that these points determine. Chord AB is a diameter precisely when $\overset{\frown}{AB} \overset{\circ}{=} 180°$, or equivalently, when $\angle C = 90°$. This completes the proof. ∎

Perhaps we should comment on the phrase "if and only if," which appears in the statement of Corollary 1.22. The if part of the statement asserts that *if AB is a diameter, then ∠C is a right angle*. In other words, this part of the corollary says that whenever we know that AB is a diameter, we can conclude that $\angle C$ is a right angle. The only if part of the corollary tells us that the *only way* that $\angle C$ can be a right angle is for AB to be a diameter. In other words, *if ∠C is a right angle, then AB must be a diameter*. The only if part of the assertion, therefore, is precisely the converse of the if part. The if and only if form of mathematical statements is so common that these four words are often combined into the single abbreviation "iff," although we do not use the abbreviation in this book.

Generally, when we are asked to prove an assertion that says something in the form "*abc* if and only if *xyz*," we are expected to provide two proofs. We prove the if part by assuming *xyz* and somehow deducing *abc*. We prove the only if part by starting all over and assuming *abc* and deducing *xyz*. Of course, it doesn't really matter whether we do if or only if first, as long as both get done. In some exceptional cases, however, it is possible to prove both the if and the only if parts of an if and only if assertion simultaneously. The proof we just gave for Corollary 1.22 is an example of this.

Here is an amusing "practical" application for Corollary 1.22. Suppose a circle is printed on a piece of paper, and we want to find its exact center. If the circle had been drawn with a compass, we could hold the paper up to the light to find the tiny hole that would mark the center, but for a printed circle, we need a different method. Take another piece of paper, with a 90° corner, and place it down over the printed circle so that it covers the approximate location of the center and so that its corner lies exactly on the circle. Now mark the point on the circle where each of the two sides of the right angled corner crosses the circle. Remove the covering paper and use a straightedge to join these two marks. By Corollary 1.22, the line segment that results is a diameter of the circle. Repeat this process to draw a second diameter and mark the point where the two diameters cross. This is clearly the center of the circle.

Observe that given any line segment AB, there is a unique circle having AB as a diameter. This, of course, is the circle centered at the midpoint of AB and having radius $\frac{1}{2}AB$. Another way to state Corollary 1.22, therefore, is that $\angle C$ is a right angle in $\triangle ABC$ if and only if point C lies on the unique circle having side AB as a diameter. Note that $\angle C < 90°$ if C lies outside of this circle and $\angle C > 90°$ if C is in the interior. Recall that angles smaller than 90° are said to be **acute**, and angles between 90° and 180° are **obtuse**.

Recall that a line is **tangent** to a circle if it meets the circle in exactly one point. Through every point P on a circle, there is a unique tangent line, which is necessarily perpendicular to the radius terminating at P. One way to see why the tangent must be perpendicular to this radius is from the point of view of calculus.

In Figure 1.30, we have drawn tangent line PT and we want to compute $\angle APT$, where AP is the diameter that extends radius OP. Choose a point Q on the circle near P and draw the secant line PQ. If we move Q closer and closer to P, we see that $\angle APT$ is the limit of $\angle APQ \overset{\circ}{=} \frac{1}{2}\,\widehat{AQ}$. But as Q approaches P, we observe that $\widehat{AQ}$ approaches $180°$ since $\widehat{AQP}$ is a semicircle. It follows that $\angle APQ$ approaches $90°$, and so AP is perpendicular to the tangent, as claimed.

Another fact that we can see in Figure 1.30 is that $\angle QPT$ between chord PQ and tangent PT is the complement of $\angle APQ$. Thus

$$\angle QPT = 90° - \angle APQ \overset{\circ}{=} 90° - \frac{1}{2}\,\widehat{AQ} \overset{\circ}{=} \frac{1}{2}(180° - \widehat{AQ}) \overset{\circ}{=} \frac{1}{2}\,\widehat{PQ}.$$

In other words, we have the following.

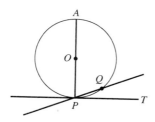

Figure 1.30

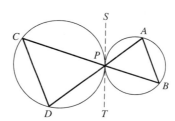

Figure 1.31

(1.23) THEOREM. *The angle between a chord and the tangent at one of its endpoints is equal in degrees to half the subtended arc.* ∎

We leave as an exercise the following consequence.

(1.24) COROLLARY. *The angle between a secant and a tangent meeting at a point outside a circle is equal in degrees to half the difference of the subtended arcs.*

Two circles are said to be **mutually tangent** at a point P if P lies on both circles and the same line through P is tangent to both circles. This can happen externally, when the two circles are on opposite sides of the tangent line, or internally, when the circles are on the same side of the tangent and one circle is inside the other.

(1.25) PROBLEM. Given two externally mutually tangent circles with common point P, draw two common secants AD and BC through P, as in Figure 1.31. Show that AB and CD are parallel.

Solution. Since BC is a transversal to the two lines AB and CD, it suffices to show that the alternate interior angles $\angle B$ and $\angle C$ are equal. Draw the common tangent ST and note that $\angle DPT \overset{\circ}{=} \frac{1}{2}\,\widehat{PD}$ by Theorem 1.23. Also, of course, $\angle C \overset{\circ}{=} \frac{1}{2}\,\widehat{PD}$, and so it follows that $\angle C = \angle DPT$. Similarly, $\angle B = \angle APS$. But $\angle DPT$ and $\angle APS$ are vertical angles, so they are equal, and it follows that $\angle C = \angle B$, as desired.   ∎

Exercises 1F

1F.1 Suppose that AB and CD are chords on two circles with equal radii. Show that $AB = CD$ if and only if $\overset{\circ}{\overparen{AB} \doteq \overparen{CD}}$.

NOTE: The assertion of this problem is really pretty obvious, but nevertheless, you should provide a proof here.

1F.2 Let A, B, C, and D be placed consecutively on a circle. Let W, X, Y, and Z be the midpoints of $\overparen{AB}$, $\overparen{BC}$, $\overparen{CD}$, and $\overparen{DA}$, respectively. Show that chords WY and XZ are perpendicular.

1F.3 Let P be a point exterior to a circle centered at point O and draw the two tangents to the circle from P. Let S and T be the two points of tangency. Show that OP bisects $\angle SPT$ and $PS = PT$.

1F.4 In the situation of the previous exercise, show that $\overparen{ST} \overset{\circ}{=} 180° - \angle SPT$.

1F.5 In $\triangle ABC$, prove that $\angle A$ is a right angle if and only if the length of the median from A to BC is exactly half the length of side BC.

1F.6 In quadrilateral $ABCD$, assume that $\angle A = 90° = \angle C$. Draw diagonals AC and BD and show that $\angle DAC = \angle DBC$.

1F.7 In the situation of the previous exercise, assume that diagonal AC bisects diagonal BD. Prove that the quadrilateral is a rectangle.

1F.8 In Figure 1.32, two circles meet at points P and Q, and diameters PA and PB are drawn. Show that line AB goes through point Q.

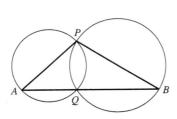

Figure 1.32

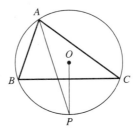

Figure 1.33

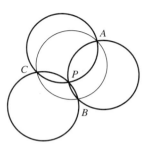

Figure 1.34

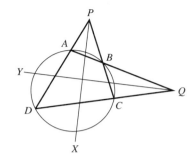

Figure 1.35

1F.9 In Figure 1.33, point O is the center of the circumcircle of $\triangle ABC$, and the bisector of $\angle A$ is extended to meet the circle at P. Prove that radius OP is perpendicular to BC.

1F.10 Given $\triangle ABC$ inscribed in a circle, draw the bisector b of the exterior angle at A. Suppose that line b is not tangent to the circle and let P be the point other than A where b meets the circle. Show that P is equidistant from B and C. In the exceptional case where b is tangent to the circle, show that A is equidistant from B and C.

1F.11 In Figure 1.34, we have drawn three circles of equal radius that go through a common point P, and we have designated by A, B, and C the three other points where these circles cross. Show that the unique circle through A, B, and C has the same radius as the original three circles.

   HINT: Use Exercise 1D.12 to show that $\triangle ABC$ is congruent to the triangle formed by the centers of the three given circles. Use the fact that circumcircles of congruent triangles have equal radii.

1F.12 In Figure 1.35, we have selected two points P and Q outside of a circle, and we have drawn two secants through each point in such a way that these four secants intersect the circle in the four points A, B, C, and D, as shown. Show that the angle bisectors PX and QY of $\angle P$ and $\angle Q$ are perpendicular to each other.

   HINT: Let U and V be the points where bisector QY meets secants PD and PC, respectively. Show that $\triangle PUV$ is isosceles by proving that $\angle PUV = \angle PVU$.

1F.13 Prove Corollary 1.24. Show that if point P is external to a circle and tangent PT and secant PAB are drawn where T, A, and B lie on the circle, then $\angle TPA \stackrel{\circ}{=} \frac{1}{2}(\overparen{BT} - \overparen{AT})$.

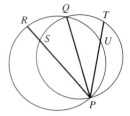

Figure 1.36

1F.14 In Figure 1.36, segment PQ is a chord common to two circles and it bisects $\angle RPT$, where R and T lie on the circles, as shown. Each of the chords PR and PT is cut by the other circle at points S and U. Prove that $RS = TU$.

1G   Polygons in Circles

A polygon is said to be **regular** if all of its sides are equal and also all of its angles are equal. An equilateral triangle, for example, certainly has equal sides, by definition, and by two applications of the pons asinorum, it is easy to see that all three angles must be equal too. An equilateral triangle, therefore, is a regular 3-gon. An equilateral 4-gon is a rhombus, but of course, a rhombus need not have equal angles, and so it is not necessarily a regular polygon. A square, however, is a regular 4-gon.

In general, for $n \geq 3$, a regular n-gon can be drawn by marking n equally spaced points around a circle and then drawing the n chords connecting consecutive points. (By equally spaced, we mean that the n arcs formed by pairs of consecutive points are all equal in degrees.) It follows by Exercise 1F.1 that the n chords are equal in length. To see that the polygon formed by the n equally spaced points is regular, we must also establish that the n angles are all equal. Each of the n arcs is clearly equal in degrees to $360/n$ degrees, and each angle of the polygon subtends an arc consisting of $n - 2$ of these small arcs. It follows by Theorem 1.16 that each of these angles is equal to $\frac{1}{2}(n - 2)(360/n) = 180(n - 2)/n$ degrees, and thus the polygon is regular, as desired. Note that the sum of all the angles of our regular n-gon is n times $180(n - 2)/n$. This is $180(n - 2)$ degrees, which fortunately agrees with what we know to be the sum of the angles of an arbitrary n-gon.

Now draw the n radii joining the center of the circle to the n equally spaced points. This subdivides the interior of the n-gon into n isosceles triangles with equal bases of length s, the common side length of the polygon. These n triangles are all congruent by SSS, and thus the lengths of the altitudes of these triangles (drawn from the center of the circle) are all equal. Any one of these altitudes is said to be an **apothegm** of the regular polygon, and we write a to denote their common length. Since the area of each of the isosceles triangles is $\frac{1}{2}sa$, we see that the area of the entire regular n-gon is $\frac{1}{2}nsa = \frac{1}{2}pa$, where we have written $p = ns$, the perimeter of the polygon.

(1.26) PROBLEM.   Fix an integer $n \geq 3$. Given a circle, how should n points on this circle be chosen so as to maximize the area of the corresponding n-gon?

It is often the case that there is symmetry in the solution to an optimization ("max-min") problem, and so it seems natural to guess that the area-maximizing inscribed polygon should be regular. In other words, the points should be equally spaced around the circle. As we shall see, this is correct.

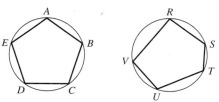

In Figure 1.37, for instance, we illustrate the case where $n = 5$. In the left diagram, $\overset{\frown}{AB}, \overset{\frown}{BC}, \overset{\frown}{CD}, \overset{\frown}{DE}$, and $\overset{\frown}{EA}$ are all equal, and thus each is $72°$. The circle in the right diagram has an equal radius, and we have placed five points around it in such a way that not all of the arcs are equal. We need to show that the area of the regular pentagon $ABCDE$ is strictly greater than that of pentagon $RSTUV$.

Figure 1.37

We begin with a discussion of general strategies for solving geometric optimization problems, as illustrated in this case. There are at least two ways we might proceed. The straightforward approach would be to show directly that the area of pentagon $ABCDE$ exceeds that of any other pentagon (such as $RSTUV$) inscribed in the same or an equal circle and which does not have all arcs equal. Alternatively, we could show that given any pentagon (such as $RSTUV$) in which the arcs are not all equal, it is possible to find a pentagon larger in area inscribed in the same circle. This would show that no pentagon other than one with equal arcs could maximize the area. If we somehow knew that an area-maximizing pentagon necessarily exists, it would follow that all of its arcs must be equal, and the problem would be solved.

For Problem 1.26, the direct approach seems difficult, but it is not hard to show that given any n-gon inscribed in a circle and having arcs that are not all equal, there exists another n-gon with larger area inscribed in the same circle. To see this, observe that since the arcs are not all equal, we can surely find two consecutive unequal arcs, and hence we can find three consecutive vertices R, S, and T of our n-gon where $\overset{\frown}{RS}$ and $\overset{\frown}{ST}$ are unequal. We show that it is possible to move point S, leaving all of the remaining $n - 1$ points fixed, so that the area is increased.

The area of the whole polygon can be viewed as the area of $\triangle RST$ plus the area of that part of the polygon that lies on the other side of chord RT. We are assuming that S is not the midpoint of $\overset{\frown}{RST}$, and we label the midpoint S', as in Figure 1.38.

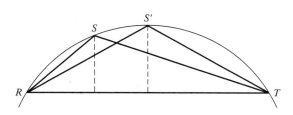

Figure 1.38

It is clear from the diagram that the perpendicular distance h from S to line RT is less than the distance h' from S' to RT. As h is the height of $\triangle RST$ with respect to the base RT and h' is the height of $\triangle RS'T$ with respect to the same base, it follows that the area of $\triangle RST$ is less than that of $\triangle RS'T$. If we move point S to S', the effect is to increase the area of the triangle without affecting the rest of the polygon, and the effect on the area of the whole polygon is thus an increase. This shows that an n-gon inscribed in a circle where the arcs are unequal cannot have the maximum possible area.

To complete the solution of Problem 1.26, it suffices now to show that among all possible n-gons inscribed in a given circle, there is one for which the area is a maximum. This maximizing polygon is necessarily regular by the foregoing argument. The existence of an area-maximizing n-gon inscribed in a circle follows from a general principle called compactness, which we explain somewhat informally and without proof.

Perhaps readers will recall from their study of calculus that if $f(x)$ is a function of a real variable x, defined for $a \leq x \leq b$, and $f(x)$ is continuous in this interval, then the function necessarily takes on a maximum value at some point c in the interval. There are two crucial hypotheses here: that f is continuous and that x runs over a closed interval. (The interval consisting of those real numbers x such that $a \leq x \leq b$ is said to be **closed** because it includes both endpoints.) More generally, it is true that a continuous function, even of several variables, takes on a maximum value if each variable runs over a closed and bounded set. (We shall define these terms presently.) In our geometry problem, we can think of the area of an n-gon inscribed in a circle as a continuous function of n variables: the n-points, which we do not require to be all different. That this function is continuous means essentially that a small perturbation in the locations of the points results in at most a small change in the area. Each point is required to lie on our circle, which as we shall see, is closed and bounded. It follows that for some choice of n points on the circle, the area function takes on a maximum value.

A set of points is **bounded** if it is contained in the interior of some circle, possibly a very large circle. Some examples of bounded sets are a line segment, a circle, and the interior of a circle; unbounded sets include a line, the exterior of a circle, and the interior of an angle.

Before we can explain what it means for an arbitrary set of points to be closed, we need another definition. Given a set $\mathcal{S}$ of points in the plane, we say that a point P is **adjacent** to $\mathcal{S}$ if every circle centered at P, no matter how small, contains at least one point of $\mathcal{S}$ in its interior. It is obvious that if P is actually a member of $\mathcal{S}$, then P is adjacent to $\mathcal{S}$, but it is also possible for a point to be adjacent to a set without actually being in the set. An endpoint of a line segment, for example, is not in the interior of the segment, but it is adjacent to the interior.

A plane set is **closed** if every point adjacent to the set is actually a member of the set. For example, the interior of a circle is not a closed set because the points of the circle are adjacent and yet are not in the set, which consists only of interior points. The disk formed by a circle together with its interior is a closed set, however, as is the circle itself. A line is a closed set and so is a line segment provided that we include the endpoints of the segment. A line segment without its endpoints, however, is not closed, nor is a circle with one point deleted.

A set that is both closed and bounded is said to be **compact**. If we are willing to believe the theorem that a real-valued continuous function of several variables, each of which runs over a compact set, attains a maximum and also a minimum value, then Problem 1.26 is solved. This is because the domain of choice for each point is a circle, which is a compact set. It is vital that the domain of choice for each point be the entire circle, with no restriction. We must not insist, therefore, that our n points all be different. Observe that the max-min existence theorem also guarantees that a minimum area can be obtained by a suitable choice of n points on a circle. The minimum area of zero is attained, for example, when all n points are the same. There are, of course, also other configurations that yield this minimum area.

Our discussion of circles would not be complete without some mention of the amazing number $\pi = 3.1415\ldots$. By definition, π is the ratio of the perimeter, also called the circumference, of a circle to its diameter. It is not amazing that this ratio is the same for all circles, independent of size. If we change the scale and multiply the diameter by some constant k, it is clear that the perimeter is also multiplied by k and the ratio remains unchanged. What does seem remarkable is that this same number π is also involved in the formula $K = \pi r^2$ giving the area of a circle in terms of its radius r. It seems that this calls for an explanation.

Fix a circle of radius r and let K_n denote the area of a regular n-gon inscribed in the circle. It should be reasonably clear, and we will not give a formal proof, that the area K of the circle is the limit of the polygon areas K_n as $n \to \infty$. We have seen that $K_n = \frac{1}{2}p_n a_n$, where p_n and a_n are, respectively, the perimeter and apothegm of a regular n-gon inscribed in our given circle. Observe that as n gets large, p_n approaches the circumference $c = 2\pi r$ of the circle and a_n approaches the radius r. Thus

$$K = \lim_{n\to\infty} K_n = \frac{1}{2}(\lim_{n\to\infty} p_n)(\lim_{n\to\infty} a_n) = \frac{1}{2}(2\pi r)(r) = \pi r^2,$$

as we expected. We mention that the formulas for the surface area and volume of a sphere in terms of its radius also involve the seemingly ubiquitous number π. Without giving proofs, we remind the reader that these formulas are $S = 4\pi r^2$ and $V = \frac{4}{3}\pi r^3$.

We cannot resist making a few more comments about the sometimes misunderstood number π. We wrote earlier that $\pi = 3.1415\ldots$, where the dots represent an infinite number of omitted decimal places. There is certainly nothing mysterious or unusual in the fact that decimal expansion of π does not terminate; the same can be said of such well-understood numbers as $1/3$ or $2/7$. These are **rational** numbers, which means that they are quotients of integers. It is well known that the decimal expansions of rational numbers either terminate or eventually repeat, but the number π is irrational, and its decimal expansion never repeats. The same can be said of numbers such as $\sqrt{2} = 1.4142\ldots$, but in a certain sense, π is even more unlike most of the numbers we meet every day than is $\sqrt{2}$.

To understand the true nature of π, we must distinguish between algebraic and transcendental numbers. Recall that a **polynomial** is an expression of the form $f(x) = a_n x^n + a_{n-1}x^{n-1} + \cdots + a_1 x + a_0$, where the constants a_i are called the **coefficients** of the polynomial $f(x)$ and we assume that the coefficient a_n of the highest power of x is nonzero. A number r is said to be a **root** of the polynomial $f(x)$ if we get 0 when we plug in r in place of x. Thus the number $r = \sqrt{2}$ is a root of the polynomial $f(x) = x^2 - 2$. Note that the coefficients of this polynomial are $a_2 = 1$, $a_1 = 0$, and $a_0 = -2$; in particular, all the coefficients are integers. A number r is said to be **algebraic** if it is a root of some polynomial with integer coefficients. Thus $\sqrt{2}$ is algebraic and so too is every rational number. For example, $2/7$ is a root of the polynomial $f(x) = 7x - 2$.

A number is **transcendental** if it is not algebraic, which means that it is not a root of any polynomial with integer coefficients. In fact, π is transcendental, and this may be what makes this number seem so mysterious and unusual. (Actually, transcendental numbers are not really unusual; there is a sense in which it is true to say that most numbers are transcendental. What is unusual is to have in hand a particular number such as π or $e = 2.7182\ldots$ that is actually known to be transcendental.) It is the fact that π is transcendental that explains why it is impossible to square a circle. We shall explain what that means in Chapter 6.

The fact that there is no polynomial equation for π does not, of course, mean that there is anything hazy or ambiguous about this number. There are, in fact, many formulas that can be proved to give π exactly. To mention just three of these, we have

$$\pi = 4 \int_0^1 \frac{dx}{1 + x^2} \qquad \pi = \left(6 \sum_{n=1}^{\infty} \frac{1}{n^2} \right)^{1/2} \qquad \pi = 4 \sum_{n=0}^{\infty} \frac{(-1)^n}{2n + 1}.$$

Exercises 1G

1G.1 In Figure 1.38, show that as S moves along $\overleftrightarrow{RT}$, the distance h from S to RT varies in direct proportion to the product $RS \cdot ST$.

HINT: Use areas.

1G.2 Suppose that A, B, C, and D are four consecutive vertices of a regular polygon but do not assume that this polygon is inscribed in a circle. We know that there is some unique circle through points A, B, C. Prove that this circle also goes through D.

HINT: Let O be the center of the circle so that $OA = OB = OC$. Use congruent triangles to show that $OC = OD$.

NOTE: It follows easily from the result of this exercise that every regular polygon can be inscribed in a circle.

1G.3 In Figure 1.39, equilateral $\triangle ABC$ is inscribed in a circle and point P is chosen arbitrarily on $\overparen{AC}$. Show that $AP + PC = PB$.

HINT: Extend chord PC to point Q, as shown, so that $PQ = PB$ and then draw BQ.

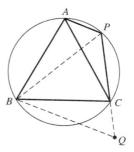

Figure 1.39

1G.4 Let K_n and a_n denote, respectively, the area and the apothegm of a regular n-gon inscribed in a circle of radius r. Show that

$$\frac{K_{2n}}{K_n} = \frac{r}{a_n}.$$

1G.5 In a circle of radius 1, show that $a_{2n} = \sqrt{(1 + a_n)/2}$, where, as in the previous exercise, a_n is the apothegm of a regular n-gon inscribed in the circle. Deduce that the following simple iterative algorithm can be used to compute an approximation to π. Start with numbers $u = 2$ and $v = 1/\sqrt{2}$. At each step, replace u by u/v and replace v by $\sqrt{(1 + v)/2}$. Show that the limiting value for u is π.

NOTE: It is not very hard to prove that at each step of this algorithm, the "error" $\pi - u$ is actually less than half of what it was at the previous step, and so it does not take very many iterations to get a fairly decent approximation to π. Try it on a computer or programmable calculator!

1H   Similarity

Informally speaking, we say that two geometric figures are similar if they have the same shape. If the figures are actually congruent, they have both the same shape and the same size, and so similarity is a weaker condition than congruence. Although we have not defined shape precisely, let us agree that the shape of a triangle is determined by its angles. This motivates our "official" definition of similarity for triangles: Two triangles are **similar** if the three angles of one are equal to the three angles of the other. If, for example, we are given $\triangle ABC$ and $\triangle XYZ$ and we know that $\angle A = \angle X$, $\angle B = \angle Y$, and $\angle C = \angle Z$, then the two triangles are similar and we write $\triangle ABC \sim \triangle XYZ$. As was the case for congruent triangles, the order in which we write vertices is significant here; it is the corresponding angles that are equal.

For figures other than triangles, the angles do not necessarily determine the shape. Two rectangles, for example, agree in all four angles, and yet one may be a square and the other not. As we shall see, it is possible to give an alternative definition of similarity that applies to all sorts of geometric figures besides triangles, but in general, one must know more than the equality of corresponding angles to establish that two figures are similar. For triangles, on the other hand, it is not even necessary to check all three pairs of angles. If two angles of one triangle are equal to the two corresponding angles of another triangle, then we automatically have equality of the other pair of angles.

(1.27) THEOREM.   *Given $\triangle ABC$ and $\triangle XYZ$, suppose $\angle A = \angle X$ and $\angle B = \angle Y$. Then $\angle C = \angle Z$, and so $\triangle ABC \sim \triangle XYZ$.*

Proof.   Since the sum of the angles of a triangle is $180°$, we have

$$\angle C = 180° - (\angle A + \angle B) = 180° - (\angle X + \angle Y) = \angle Z,$$

as required.   ∎

When we use Theorem 1.27 to prove that two triangles are similar, we say that the triangles are similar by AA. Of course, AA is an abbreviation for "angle-angle." But what is the use of knowing that two triangles are similar? What can we deduce from similarity?

(1.28) THEOREM.   *If $\triangle ABC \sim \triangle XYZ$, then the lengths of the corresponding sides of these two triangles are proportional.*

Before we prove Theorem 1.28, which is of fundamental importance, let us be sure that we understand precisely what it is telling us. The assertion that the corresponding sides are proportional tells us that there is some positive number λ, called a **scale factor**, such that

$$AB = \lambda \cdot XY,$$
$$BC = \lambda \cdot YZ, \quad \text{and}$$
$$AC = \lambda \cdot XZ.$$

For example, if the scale factor λ is equal to 3, this would say that each side of $\triangle ABC$ is three times as long as the corresponding side of $\triangle XYZ$, and if $\lambda = 1/2$, this would say that the sides of $\triangle ABC$ are half as long as the corresponding sides of $\triangle XYZ$. If the sides of $\triangle ABC$ are 3 km, 4 km, and 5 km and the corresponding sides of $\triangle XYZ$ are 3 cm, 4 cm, and 5 cm, then the sides of these two triangles are proportional with a scale factor $\lambda = 100,000$. Observe that proportionality of side lengths is symmetric: If the sides of $\triangle ABC$ are proportional to the sides of $\triangle XYZ$ with scale factor λ, then the sides of $\triangle XYZ$ are proportional to those of $\triangle ABC$ with scale factor $1/\lambda$.

There is another way to think about proportionality. Suppose we know that $\triangle ABC \sim \triangle XYZ$. Theorem 1.28 tells us that the side lengths of $\triangle ABC$ are proportional to those of

$\triangle XYZ$, but it does not mention any particular scale factor. We can determine λ, however, from any one of the equations

$$\lambda = \frac{AB}{XY},$$

$$\lambda = \frac{BC}{YZ}, \quad \text{or}$$

$$\lambda = \frac{AC}{XZ}.$$

The significance of proportionality is that we get the same value for λ from each of these equations. In other words, the corresponding sides are proportional if and only if

$$\frac{AB}{XY} = \frac{BC}{YZ} = \frac{AC}{XZ}.$$

The key to the proof of Theorem 1.28 is the following lemma, which relates parallelism to proportionality.

(1.29) LEMMA. *Let U and V be points on sides AB and AC of $\triangle ABC$. Then $UV \| BC$ if and only if*

$$\frac{AU}{AB} = \frac{AV}{AC}.$$

Proof.   Write $\alpha = UB/AB$ and $\beta = VC/AC$. Then $AU = AB - UB = (1 - \alpha)AB$, and similarly, $AV = (1 - \beta)AC$. Thus

$$\frac{AU}{AB} = 1 - \alpha \quad \text{and} \quad \frac{AV}{AC} = 1 - \beta.$$

It follows that the ratios AU/AB and AV/AC are equal if and only if $\alpha = \beta$. It suffices now to show that α and β are equal if and only if $UV \| BC$.

Draw CU, as in Figure 1.40, and compare the area of $\triangle BUC$ with that of $\triangle ABC$. Viewing AB and UB as bases, we see that these triangles have equal heights, and thus

$$\frac{K_{BUC}}{K_{ABC}} = \frac{UB}{AB} = \alpha$$

and $K_{BUC} = \alpha K_{ABC}$. Similarly, $K_{BVC} = \beta K_{ABC}$ and it follows that the quantities α and β are equal if and only if $\triangle BUC$ and $\triangle BVC$ have equal areas. We therefore need to show that this equality of areas happens if and only if $UV \| BC$.

Since $\triangle BUC$ and $\triangle BVC$ share base BC, their areas are equal if and only if they have equal heights UE and VF. Our task, therefore, is to show that $UE = VF$ if and only if UV and EF are parallel. Observe that UE and VF are parallel since each of these lines is perpendicular to BC. If $UE = VF$, therefore, it follows by Theorem 1.8 that $UVFE$ is a parallelogram, and thus $UV \| EF$, as required, as in

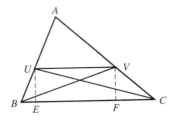

Figure 1.40

Figure 1.40. Conversely, if $UV \parallel EF$, then $UVEF$ is a parallelogram by definition, and so $UE = VF$ by Theorem 1.6. This completes the proof.   ∎

Proof of Theorem 1.28.   Recall that we are given that $\triangle ABC \sim \triangle XYZ$, and we will prove that $XY/AB = XZ/AC$. Similar reasoning would also show that $XY/AB = YZ/BC$, and thus all three ratios are equal and the sides are proportional.

If $XY = AB$, then $\triangle ABC \cong \triangle XYZ$ by ASA. In this case, $XZ = AC$, and so $XY/AB = 1 = XZ/AC$, as desired. We can thus suppose that XY and AB are unequal, and it is no loss to assume that XY is the shorter of these two segments. Choose point U on side AB of $\triangle ABC$ so that $AU = XY$ and draw UV parallel to BC, where V lies on side AC. Since $UV \parallel BC$, we have $\angle AUV = \angle B = \angle Y$ and $\angle AVU = \angle C = \angle Z$. But $AU = XY$, and so we deduce that $\triangle AUV \cong \triangle XYZ$ by SAA. In particular, we have $AV = XZ$.

Since $UV \parallel BC$, we know by Lemma 1.29 that $AU/AB = AV/AC$. Since $AU = XY$ and $AV = XZ$, it follows that $XY/AB = XZ/AC$, and the proof is complete.   ∎

What is wrong with the following easy "proof" of Theorem 1.28? In $\triangle ABC$, the law of sines gives $a/\sin(A) = b/\sin(B)$, and thus $a/b = \sin(A)/\sin(B)$. In any triangle, therefore, the ratio of two sides is equal to the ratio of the sines of the opposite angles. In the situation of Theorem 1.28, we have

$$\frac{BC}{AC} = \frac{\sin(A)}{\sin(B)} = \frac{\sin(X)}{\sin(Y)} = \frac{YZ}{XZ}$$

and a bit of algebra yields $XZ/AC = YZ/BC$. The equality $XZ/AC = XY/AB$ follows similarly, and thus the sides of the triangles are proportional, as desired, and the proof is complete.

To see why this argument is not valid, recall that in proving the law of sines, we used the fact that if $\triangle ABC$ is a right triangle with $\angle C = 90°$ and $\angle A = \theta$, then $\sin(\theta) = BC/AB$. Underlying this "opposite-over-hypotenuse" formula for the sine of an angle is the assumption that the ratio BC/AB has a value that is independent of the particular triangle in question. In other words, if $\triangle XYZ$ is another triangle and $\angle Z = 90°$ and $\angle X = \theta$, we expect that it is also true that $\sin(\theta) = YZ/XY$. This is correct since $\triangle ABC \sim \triangle XYZ$ by AA, and hence by Theorem 1.28, corresponding sides are proportional and we have $YZ = \lambda BC$ and $XY = \lambda AB$ for some scale factor λ.

It follows that $YZ/XY = BC/AB$, as expected. We need Theorem 1.28 to justify the opposite-over-hypotenuse formula for the sine, and hence it would be circular reasoning to use sines to prove Theorem 1.28. Of course, now that we have proved Theorem 1.28, we can safely use sines and other trigonometric functions to study triangles.

As a demonstration of the power of similar triangles to prove interesting theorems, we present the following.

(1.30) PROBLEM.   In $\triangle ABC$, draw UV parallel to BC as in Figure 1.41. Show that the intersection of BV and UC lies on the median of $\triangle ABC$ from point A.

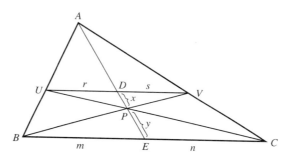

Figure 1.41

Solution.   As shown in Figure 1.41, let P be the intersection point of BV and UC. Draw line AP and let D and E be the points where this line crosses UV and BC. For convenience, we denote the lengths UD, DV, BE, EC, DP, and PE by r, s, m, n, x, and y, respectively, as indicated in the figure. Our task is to prove that E is the midpoint of AC, and so we need to show that $m = n$.

First, note that $\triangle AUD \sim \triangle ABE$ by AA because $UD \| BE$, and so $\angle AUD = \angle ABE$ and $\angle ADU = \angle AEB$. We conclude that $r/m = AD/AE$, and similarly, $s/n = AD/AE$. This gives $r/m = s/n$, and hence $n/m = s/r$.

Now consider $\triangle DUP$ and $\triangle ECP$. Because $UD \| EC$, we have $\angle DUP = \angle PCE$ and $\angle UDP = \angle CEP$. Thus $\triangle DUP \sim \triangle ECP$, and hence $r/n = x/y$. Similar reasoning with $\triangle DVP$ and $\triangle EBP$ yields $s/m = x/y$, and we conclude that $r/n = s/m$, and thus $m/n = s/r$. Since we previously had $n/m = s/r$, we see that $m/n = n/m$, and it follows that $n = m$, as required.   ■

There is a case of Lemma 1.29 that occurs sufficiently often to deserve separate mention.

(1.31) COROLLARY.   *Let U and V be the midpoints of sides AB and AC, respectively, in $\triangle ABC$. Then $UV \| BC$ and $UV = \frac{1}{2}BC$.*

Proof.   That $UV \| BC$ is immediate from Lemma 1.29 since $AU/AB = 1/2 = AV/AC$. In this situation, $\triangle AUV \sim \triangle ABC$ by AA, and thus by Theorem 1.28, we know that $UV/BC = AU/AB = 1/2$. It follows that $UV = \frac{1}{2}BC$, as required.   ■

There are two useful similarity criteria other than AA. These are SSS and SAS, which should not be confused with the congruence criteria having the same names. The SSS similarity criterion asserts that two triangles must be similar if we know that the three sides of one of the triangles are proportional to the three corresponding sides of the other triangle. We state this formally as follows.

(1.32) THEOREM.   *Suppose that the sides of $\triangle ABC$ are proportional to the corresponding sides of $\triangle XYZ$. Then $\triangle ABC \sim \triangle XYZ$.*

Proof.   By hypothesis, each of XY, XZ, and YZ is equal, respectively, to some scale factor λ times AB, AC, and BC. If it happens that $AB = XY$, then $\lambda = 1$ and all three sides of $\triangle ABC$ are equal to the corresponding sides of $\triangle XYZ$. In this case, the triangles are congruent by SSS, and so they are automatically similar. We can thus assume that AB and XY are unequal, and it is no loss to assume that XY is the shorter of these two segments.

Choose point U on segment AB such that $AU = XY$ and draw UV parallel to BC with V on AC. It follows that $\triangle AUV \sim \triangle ABC$ by AA, and hence the corresponding sides of these two triangles are proportional. The scale factor for this proportionality is $AU/AB = XY/AB = \lambda$, and thus $AV = \lambda AC = XZ$ and $UV = \lambda BC = YZ$. It follows that $\triangle AUV \cong \triangle XYZ$ by SSS, and hence $\angle AUV = \angle Y$ and $\angle AVU = \angle Z$. Thus $\angle B = \angle Y$ and $\angle C = \angle Z$, and hence $\triangle ABC \sim \triangle XYZ$ by AA.   ∎

One way to appreciate the significance of the SSS similarity criterion in Theorem 1.32 is to imagine a photocopy machine with a control that enables the user to make a reduced-size image. If the reducing control is set to 75%, for example, then the copy will be only three quarters as large as the original, but in other respects, it resembles the original document. Suppose that the original happens to be a geometry diagram, such as those in this book. (Of course, owners of this book would not really allow unauthorized reproduction in violation of copyright rules, but this is just a hypothetical example.) Lengths of line segments in the image are shorter than the corresponding lengths in the original figure, and they are proportional with a scale factor of $\lambda = .75$. But how do the angles in the image compare with the corresponding angles in the original? We know, of course, that the image has the same shape (whatever that means) as the original, and so we expect every angle in the image to equal its corresponding angle in the original. Theorem 1.32 justifies this expectation, as follows.

Suppose the original figure contains some angle (say, $\angle ABC$) and suppose that the points in the reduced-size copy that correspond to A, B, and C, are X, Y, and Z, respectively. To see why $\angle ABC = \angle XYZ$, we can suppose that segment AC was drawn in the original so that segment XZ appears in the copy. For simplicity, we shall assume that neither $\angle ABC$ nor $\angle XYZ$ is a straight angle so that we have actual triangles $\triangle ABC$ in the original and $\triangle XYZ$ in the photocopy. We know that the sides of $\triangle XYZ$ are proportional to the sides of $\triangle ABC$, with scale factor $\lambda = .75$, and thus $\triangle ABC \sim \triangle XYZ$ by SSS. In particular, $\angle ABC = \angle XYZ$, as expected.

We are now ready to define similarity for figures that are not necessarily triangles. Two arbitrary figures are **similar** if for each point in one of them, there is a corresponding point in the other so that all distances in the first figure are proportional, with some particular scale factor λ, to the corresponding distances in the second figure. Caution: This is not how we defined similarity for triangles, but we know by Theorem 1.28 that similar triangles do satisfy this condition, and conversely, by Theorem 1.32, we see that triangles that are similar in this proportionality sense are actually similar triangles, according to our earlier definition.

Finally, we come to the SAS similarity criterion, which guarantees that two triangles are similar if two pairs of corresponding sides are proportional and the included angles are equal.

(1.33) THEOREM. *Given $\triangle ABC$ and $\triangle XYZ$, assume that $\angle X = \angle A$ and that $XY/AB = XZ/AC$. Then $\triangle ABC \sim \triangle XYZ$.*

Proof. If $XY = AB$, then $XZ = AC$ and the triangles are congruent by SAS, and hence they are automatically similar. As in the proof of Theorem 1.32, therefore, we can assume that $XY < AB$, and we choose point U on side AB of $\triangle ABC$ so that $AU = XY$. As in the previous proof, we draw UV parallel to BC with V on side AC, and we observe that $\triangle AUV \sim \triangle ABC$ by AA. Then $AV/AC = AU/AB = XY/AB = XZ/AC$, and hence $AV = XZ$. We conclude that $\triangle AUV \cong \triangle XYZ$ by SAS, and thus $\angle XYZ = \angle AUV = \angle ABC$. It now follows by AA that $\triangle ABC \sim \triangle XYZ$, as required. ∎

A striking and nontrivial application of Theorem 1.33 is the following.

(1.34) PROBLEM. In Figure 1.42, point P lies outside of parallelogram $ABCD$ and $\angle PAB = \angle PCB$. Show that $\angle APD = \angle CPB$.

Although we mentioned that the key to this problem is the SAS similarity criterion, there seem to be few similar triangles in Figure 1.42. If we believe the assertion of the problem, however, then $\triangle PCB \sim \triangle PAX$, where X is the unlabeled point of the diagram where PD crosses AB. If we could somehow prove that these triangles actually are similar, then the equality of $\angle APD$ and $\angle CPB$ would follow. But it is not clear how we can possibly use the SAS criterion to prove that $\triangle PCB \sim \triangle PAX$.

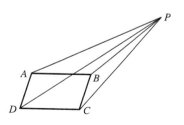

Figure 1.42

   The secret to solving this problem is to draw a number of additional lines. Extend sides CD and AD of the given parallelogram so that they meet lines PA and PC at E and F, respectively, and draw segments AC and EF, as shown in Figure 1.43. We now have an embarrassment of riches—there are so many pairs of similar triangles in this figure that it is still not quite obvious how we should proceed.

Solution to Problem 1.34.   Observe that $\angle DEA = \angle BAP$ because these are corresponding angles for the parallel lines ED and AB. Also, we have $\angle DFC = \angle BCP$ by similar reasoning. Since $\angle BAP = \angle BCP$ by hypothesis, we conclude that $\angle DEA = \angle DFC$. Also, $\angle ADE = \angle CDF$ since these are vertical angles, and so $\triangle ADE \sim \triangle CDF$ by AA. By Theorem 1.28, we conclude that $AD/CD = ED/FD$, and hence by elementary algebra, we have $AD/ED = CD/FD$. Since we know that $\angle ADC = \angle EDF$, it follows via SAS that $\triangle ADC \sim \triangle EDF$. In particular, we have $\angle CAD = \angle FED$.

   Observe now that

$$\angle FEP = \angle FED + \angle DEA = \angle CAD + \angle BAP = \angle ACB + \angle BCP = \angle ACP .$$

where the penultimate equality holds because $\angle BAP = \angle BCP$ by hypothesis. Since $\angle EPF = \angle CPA$, we can now conclude that $\triangle EPF \sim \triangle CPA$ by AA, and so we have $PC/PE = AC/FE$. We also have $AC/FE = AD/ED$ because we know that $\triangle ADC \sim \triangle EDF$. Combining these equalities, we obtain

$$\frac{PC}{PE} = \frac{AC}{FE} = \frac{AD}{ED} = \frac{CB}{ED} ,$$

where we have used the fact that $AD = CB$ to obtain the last equality.

   We can now argue that $\triangle PCB \sim \triangle PED$. Certainly, $\angle PED = \angle PCB$ since each of these is equal to $\angle PAB$, and so the similarity follows by SAS since we

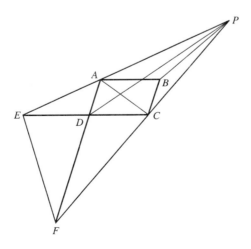

Figure 1.43

have just seen that $PC/PE = CB/ED$. We conclude that $\angle CPB = \angle EPD$, as required.   ∎

What follows is a much easier result that can be proved using similar triangles.

(1.35) THEOREM.   *Given a circle and a point P not on the circle, choose an arbitrary line through P, meeting the circle at points X and Y. Then the quantity* $PX \cdot PY$ *depends only on the point P and is independent of the choice of the line through P.*

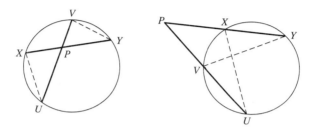

Figure 1.44

Proof.   Draw a second line through P that also meets the circle in two points, U and V, as shown in Figure 1.44. We must show that $PX \cdot PY = PU \cdot PV$. There are two cases that we must consider: where P lies inside the circle and where P lies outside.

In either case, draw line segments UX and VY, as shown in Figure 1.44, and observe that $\angle U = \angle Y$ since these inscribed angles subtend the same arc. Next, observe that $\angle XPU = \angle VPY$ since these are vertical angles when P is inside the circle, and they are in fact the same angle when P is outside the circle. In either case, therefore, we have $\triangle PXU \sim \triangle PVY$ by AA, and thus $PX/PV = PU/PY$. Elementary algebra now yields $PX \cdot PY = PU \cdot PV$, as required.   ∎

As our final application of similarity in this chapter, we offer another proof of the Pythagorean theorem. It is based on the following easy lemma.

(1.36) LEMMA.   *Suppose $\triangle ABC$ is a right triangle with hypotenuse AB and let CP be the altitude drawn to the hypotenuse. Then $\triangle ACP \sim \triangle ABC \sim \triangle CBP$.*

Proof.   The first similarity follows by AA since $\angle A = \angle A$ and $\angle APC = 90° = \angle ACB$. The proof of the second similarity is similar.   ∎

Alternative proof of Pythagoras' theorem.   In the situation of Lemma 1.36, write as usual $a = BC$, $b = AC$, and $c = AB$. Since $\triangle ACP \sim \triangle ABC$, we have $AP/AC = AC/AB$, and it follows that $AP = b^2/c$. Similarly, $BP = a^2/c$, and since $c = AB = AP + PB$, we conclude that $c = a^2/c + b^2/c$. Multiplication by c yields the equation $c^2 = a^2 + b^2$, as desired.   ∎

We close this section on similarity with a brief discussion of the areas of similar figures. Suppose we have two similar polygons $\mathcal{P}$ and $\mathcal{Q}$, where the distances in $\mathcal{Q}$ are obtained from those in $\mathcal{P}$ by multiplication by some fixed scale factor λ. If it happens that $\mathcal{P}$ and $\mathcal{Q}$ are squares, with side lengths p and q, respectively, then $q = \lambda p$, and hence $K_{\mathcal{Q}} = q^2 = \lambda^2 p^2 = \lambda^2 K_{\mathcal{P}}$. In other words, to find the area of $\mathcal{Q}$, we multiply the area of $\mathcal{P}$ by the square of the scale factor that was used for distances.

The rule "multiply by the square of the scale factor" works for arbitrary polygons and not just for squares. To see why, imagine that $\mathcal{P}$ is subdivided into a very large number of very small squares with a little left over at the edges. If $\mathcal{Q}$ is subdivided into corresponding squares, then each square in $\mathcal{Q}$ has area equal to λ^2 times the area of the corresponding square in $\mathcal{P}$. Thus the sum of the areas of the little squares that almost comprise $\mathcal{Q}$ is λ^2 times the sum of the little squares in $\mathcal{P}$, and it follows that, with vanishingly small error, we have $K_{\mathcal{Q}} = \lambda^2 K_{\mathcal{P}}$.

We do not pretend that the discussion of the previous paragraph provides a rigorous proof of the theorem that the areas of similar figures are related to each other by the square of the scale factor that relates corresponding distances. Our argument does explain why this fact should be true, however, and it can be transformed into a correct proof using limits and other ideas from calculus.

Exercises 1H

1H.1 Given $\triangle ABC$, let X, Y, and Z be the midpoints of sides BC, AC, and AB, respectively, and draw $\triangle XYZ$. Show that this divides the original triangle into four congruent triangles.

1H.2 Show that the three medians of a triangle go through a common point.

HINT: Use Problem 1.30.

NOTE: This point is called the **centroid** of the triangle. We shall have more to say about it in Chapter 2.

1H.3 Let $ABCD$ be an arbitrary quadrilateral. Show that the midpoints of the four sides are the vertices of a parallelogram.

1H.4 Given a convex quadrilateral $ABCD$, draw the two diagonals, AC and BD, and assume that $\angle ADB = \angle ACB$. Show that $\angle ABD = \angle ACD$.

NOTE: One way to prove this is to show that point C must lie on the circumcircle of $\triangle ABD$. There is also a way to prove this without circles, using similar triangles. Try to find such a proof.

1H.5 Given two points X and Y on a circle, a point P is chosen on line XY, outside of the circle, and tangent PT is drawn, where T lies on the circle. Show that $PX \cdot PY = (PT)^2$.

1H.6 In Figure 1.45, line segments PA, PB, and PC join point P to the three vertices of $\triangle ABC$. We have chosen point Y on PB and drawn YX and YZ parallel to BA and BC, respectively, where X lies on PA and Z lies on PC. Prove that $XZ \| AC$.

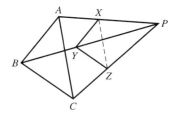

Figure 1.45

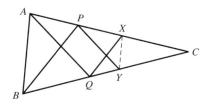

Figure 1.46

1H.7 In $\triangle ABC$, draw line segments AQ and BP, where P and Q lie on sides AC and BC, respectively. Now draw PY parallel to AQ and QX parallel to BP, where X and Y lie on AC and BC, as shown in Figure 1.46. Show that $XY\|AB$.

1H.8 Let X be a point on side BC of $\triangle ABC$ and draw AX. Let U and V be points on side AB and AC, respectively, chosen so that $UV\|BC$, and let Y be the point where AX cuts UV. Show that $UY/YV = BX/XC$.

NOTE: In particular, this exercise tells us that if AX is a median of $\triangle ABC$, then AY is a median of $\triangle AUV$.

1H.9 In Figure 1.47, we started with a circle having parallel chords AB and CD and we chose points P and Q, as shown. Let X be the intersection point of chords PB and QC and let Y be the intersection point of chords QA and PD. Show that $XY\|CD$.

HINT: Let Z be the intersection point of PD and QC and show that $\triangle PZX \sim \triangle QZY$. Use this to show that $ZX/ZC = ZY/ZD$.

1H.10 Given a circle centered at point O and an arbitrary point P, consider the locus of all points Y that occur as midpoints of segments PX, where X lies on the given circle. Show that this locus is a circle with radius half that of the original circle. Locate the center of the locus.

1H.11 A tangent PA and a secant PB are drawn to a circle from an outside point P, and the circle goes through the midpoint M of PB, as shown in Figure 1.48. If the length $AM = 1$, compute the length AB.

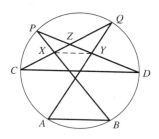

Figure 1.47

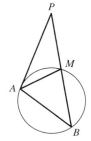

Figure 1.48

CHAPTER TWO

Triangles

2A The Circumcircle

As we saw in Theorem 1.15, there is a unique circle through any three noncollinear points. Each triangle, therefore, is inscribed in exactly one circle, called its **circumcircle**. The center and the length of the radius of this circumscribed circle are referred to as the **circumcenter** and the **circumradius** of the triangle, and the usual notation is O for the circumcenter and R for the circumradius.

Given $\triangle ABC$, we know that its circumcenter O is equidistant from vertices A and B, and so it lies on the perpendicular bisector of side AB. Similarly, O lies on the perpendicular bisectors of sides BC and AC and we have the following.

(2.1) THEOREM.   *The three perpendicular bisectors of the sides of a triangle are concurrent at the circumcenter of the triangle.* ∎

In Figure 2.1, we see that there are at least two possibilities. In the diagram on the left, the circumcenter O, where the three perpendicular bisectors meet, is an interior point of $\triangle ABC$, while on the right, O lies outside of the triangle. The explanation for this difference is that on the left $\angle A$ is acute and on the right it is obtuse. In general, given an arbitrary $\triangle ABC$, we see that if $\angle A < 90°$, then the arc of the circumcircle subtended by $\angle A$ is less than a semicircle. In this case, the center of the circle lies on the same side of line BC as does A. If all three angles are acute, similar reasoning shows that O lies on the "inside" side of all three sides of the triangle, and so it is interior to $\triangle ABC$. If the angle at some vertex (say, A) exceeds $90°$, however, then the angle subtended is more

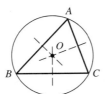 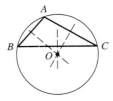

Figure 2.1

50

than a semicircle, and the center O lies on the side of line BC opposite from A. In other words, the circumcenter lies outside of the triangle in this case. Finally, if $\angle A = 90°$, then $\overset{\frown}{BC}$ is a semicircle. In this case, the circumcenter O lies on the hypotenuse BC, and thus BC is a diameter of the circumcircle. For a right triangle, therefore, the circumcenter is the midpoint of the hypotenuse and the circumradius $R = \frac{1}{2}BC$.

(2.2) PROBLEM. At which points do the perpendicular bisectors of the sides of a triangle meet the circumcircle?

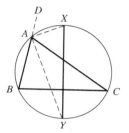

Figure 2.2

Solution. In Figure 2.2, line XY is the perpendicular bisector of side BC of $\triangle ABC$, and X and Y lie on the circumcircle. We are (rather ambiguously) asked to describe or somehow characterize the points X and Y.

Since Y lies on the perpendicular bisector of BC, we know that Y is equidistant from B and C, and thus chords BY and CY are equal. It follows that $\overset{\frown}{BY} \overset{\circ}{=} \overset{\frown}{CY}$, and thus $\angle BAY = \angle CAY$. This shows that Y is the point where the bisector of $\angle A$ in the original triangle meets the circumcircle.

It is not quite so obvious how to characterize the point X, but a study of the diagram and the symmetry of the situation leads one to guess that X is where the bisector of the *exterior* angle at A meets the circumcircle. To prove this, we extend AB to D, as shown, and we draw AX. Our goal is to show that $\angle DAX = \angle CAX$.

Note that the circumcenter of $\triangle ABC$ lies on the perpendicular bisector XY of BC, and thus XY is a diameter of the circle. It follows that $\angle XAY = 90°$, and thus

$$\angle DAX = 180° - \angle BAX = 180° - (90° + \angle BAY) = 90° - \angle BAY.$$

Since we know that $\angle BAY = \angle CAY$, this yields

$$\angle DAX = 90° - \angle CAY = \angle XAY - \angle CAY = \angle CAX,$$

as required. ∎

Recall that in Chapter 1, we used area principles to derive the law of sines:

$$\frac{a}{\sin(A)} = \frac{b}{\sin(B)} = \frac{c}{\sin(C)}.$$

This formula tells us that there is a certain mysterious quantity that can be computed in three different ways: from the perspective of each of the three vertices of $\triangle ABC$. The fact that we get the same answer by computing each of $a/\sin(A)$, $b/\sin(B)$, and $c/\sin(C)$ suggests that this quantity should mean something, but our earlier proof did not make clear what this meaning might be. In fact, the common value of the three fractions in the law of sines turns out to be exactly the diameter of the circumcircle.

(2.3) THEOREM (Extended Law of Sines).   *Given $\triangle ABC$ with circumradius R, write as usual a, b, and c to denote the lengths of the sides opposite vertices A, B, and C, respectively. Then*

$$\frac{a}{\sin(A)} = \frac{b}{\sin(B)} = \frac{c}{\sin(C)} = 2R.$$

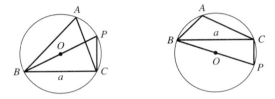

Figure 2.3

Proof.   We shall show that $a/\sin(A) = 2R$; the other equalities are proved similarly. Draw the circumcircle of $\triangle ABC$ and let BP be the diameter through point B. We have drawn two of the possible cases in Figure 2.3; there is also the possibility that $P = C$, which occurs if $\angle A$ is a right angle. We will need to treat these three situations separately.

First, assume that $\angle A < 90°$, as in the left diagram of Figure 2.3. Since BP is a diameter, we see that $\triangle PBC$ is a right triangle with hypotenuse BP of length $2R$, and thus $\sin(P) = BC/BP = a/2R$. But $\angle A = \angle P$ because they subtend the same arc; hence $\sin(A) = a/2R$, and the desired equality follows.

If $\angle A = 90°$, then $\triangle ABC$ is a right triangle and the hypotenuse BC is a diameter of the circumcircle. Thus $a = BC = 2R$, and since $\sin(A) = 1$ in this case, we have $a/\sin(A) = a = 2R$, as required.

Finally, we refer to the diagram on the right of Figure 2.3, where $\angle A > 90°$. As in the first case, we see that here too $\triangle PBC$ is a right triangle and that $\sin(P) = a/2R$. In this case, A and P are opposite vertices of an inscribed quadrilateral, and hence $\angle A = 180° - \angle P$. As we remember from trigonometry, it follows that $\sin(A) = \sin(P)$, and thus $\sin(A) = a/2R$ in this case too.   ■

Here is a nice little application of the extended law of sines. We leave to the exercises a direct, nontrigonometric proof of this fact.

(2.4) PROBLEM.   Given isosceles $\triangle ABC$, choose a point P on the base BC. Show that $\triangle ABP$ and $\triangle ACP$ have equal circumradii.

Solution.   Apply the extended law of sines at vertex B in $\triangle ABP$ to deduce that the diameter of the circumcircle of $\triangle ABP$ is $AP/\sin(B)$. Similarly, by applying the extended law of sines at vertex C in $\triangle ACP$, we see that the diameter of the circumcircle of $\triangle ACP$ is $AP/\sin(C)$. But since $\triangle ABC$ is isosceles, the pons asinorum tells us that $\angle B = \angle C$, and thus $\sin(B) = \sin(C)$ and the two diameters are equal.

■

Another application of the extended law of sines is the following formula relating the circumradius, area, and side lengths of a triangle.

(2.5) COROLLARY.   *Let R and K denote the circumradius and area of $\triangle ABC$, respectively, and let a, b, and c denote the side lengths, as usual. Then $4KR = abc$.*

Proof.   We know from Chapter 1 that $2K = ab\sin(C)$, and by the extended law of sines, we have $2R = c/\sin(C)$. Multiplication of these two equations yields the desired result.

■

We begin now to explore the connection between altitudes and circumcircles.

(2.6) PROBLEM.   Given acute angled $\triangle ABC$, draw altitude AD and note that point D must lie on line segment BC, as in Figure 2.4. Extend AD beyond D to point X on the circumcircle. Observe that $AD > DX$, and thus there is a point H on segment AD for which $HD = DX$. Show that line BH is perpendicular to AC.

Since by assumption, $\angle B$ and $\angle C$ are acute, it should be clear that point D, which is the foot of altitude AD, lies between B and C, and so it lies on line segment BC, as asserted. The fact that $AD > DX$ is, perhaps, not quite as obvious, and so we sketch an argument that explains why D must lie below the midpoint of AX in Figure 2.4. Consider BC and the perpendicular bisector of AX. These are parallel lines that cut AX at point D and at the midpoint of AX, respectively, and so it is enough to show that BC is below the perpendicular bisector. But the perpendicular bisector of AX goes through the center of the circle, and thus it suffices to observe that BC is below the center. This is true because $\angle A$ is acute, and hence it subtends an arc that is less than a semicircle.

We have now justified the assertion that the point H actually lies inside the triangle, as we have drawn it in Figure 2.4.

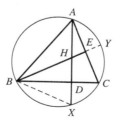

Figure 2.4

Solution to Problem 2.6.   Let E and Y be the points where BH meets side AC and where it meets the circle, respectively. Since D is the midpoint of HX, we see that BC is the perpendicular bisector of HX, and thus $BH = BX$. Since BD is an altitude of isosceles $\triangle HBX$, it is also an angle bisector, and so $\angle YBC = \angle XBC$. It follows from this that $\overset{\frown}{YC} \overset{\circ}{=} \overset{\frown}{XC}$.

By two applications of Corollary 1.19, we have

$$90° = \angle ADB \overset{\circ}{=} \frac{1}{2}(\overset{\frown}{AB} + \overset{\frown}{XC}) \overset{\circ}{=} \frac{1}{2}(\overset{\frown}{AB} + \overset{\frown}{YC}) \overset{\circ}{=} \angle AEB,$$

and thus $\angle AEB$ is a right angle, as desired.   ∎

In the situation of Problem 2.6, we know that the altitude of $\triangle ABC$ from vertex B crosses altitude AD at the specified point H. By similar reasoning, we can deduce that the altitude from vertex C also crosses AD at this same point H, and thus the three altitudes are concurrent. Although it is true for every triangle that the altitudes are concurrent at a point called the **orthocenter**, we have so far proved this striking theorem only in the case of acute angled triangles. We give a different and better proof that works in all cases in Section 2C.

By Problem 2.6, we know that the three points that result when the orthocenter of an acute angled triangle is reflected in the sides of the triangle all lie on the circumcircle. (Recall that a point Q is the **reflection** of a point P in a given line if the line is the perpendicular bisector of segment PQ.) It is natural to ask if there is any point other than the orthocenter whose reflections in the sides of the triangle all lie on the circumcircle. To see that the answer is no, consider Figure 2.5.

Start with acute angled $\triangle ABC$ and reflect each of the vertices A, B, and C in the opposite side of the triangle to obtain points U, V, and W, respectively. Next, draw the circumcircles of $\triangle BCU$, $\triangle CAV$, and $\triangle ABW$ and observe that these appear to go through a common point. To understand what is going on here, note that $\triangle UBC$ is the reflection of $\triangle ABC$ in line BC, and thus the circumcircle of $\triangle UBC$ is just the reflection of the circumcircle of $\triangle ABC$ in this line. It follows that the circumcircle of $\triangle UBC$ is the locus of all points whose reflection in line BC lies on the circumcircle of the original triangle. In particular, by Problem 2.6, we know that the orthocenter H of $\triangle ABC$ lies on this circle.

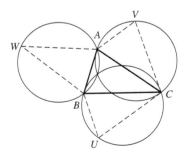

Figure 2.5

By similar reasoning, we know that H lies on each of the other two circles in Figure 2.5, and hence the three circles do indeed have a point in common. The ortho-center H lies on all three circles, and since these circles clearly cannot have more than one common point, we see that H is the only point all of whose reflections in the sides of $\triangle ABC$ lie on the circumcircle of this triangle.

We can draw a further conclusion in this situation. The three circles through point H in Figure 2.5 are clearly the circumcircles of $\triangle HBC$, $\triangle AHB$, and $\triangle ABH$. Since each of these circles is a reflection of the circumcircle of $\triangle ABC$, all four circumcircles have equal radii. This fact appears with a different proof as Corollary 2.15 in Section 2C.

Exercises 2A

2A.1 Show that quadrilateral $ABCD$ can be inscribed in a circle if and only if $\angle B$ and $\angle D$ are supplementary.

HINT: To prove if, show that D lies on the unique circle through A, B, and C.

NOTE: A quadrilateral inscribed in a circle is said to be **cyclic**.

2A.2 Given $\triangle ABC$ and $\triangle XYZ$, assume $AB = XY$ and that $\angle C$ and $\angle Z$ are supplementary. Prove without using the extended law of sines that the two triangles have equal circumradii.

HINT: Move $\triangle XYZ$ so that AB and XY coincide.

NOTE: This provides an alternative solution to Problem 2.4.

2A.3 Given acute angled $\triangle ABC$, extend the altitudes from A, B, and C to meet the circumcircle at points X, Y, and Z, respectively. Show that lines AX, BY, and CZ bisect the three angles of $\triangle XYZ$.

2A.4 In Figure 2.6, altitude AD of $\triangle ABC$ has been extended to meet the circumcircle at point X. Point H was chosen on line AD so that $HD = DX$. Show that BH is perpendicular to AC.

HINT: Let P be the point where BH crosses the circle and show that $\overset{\circ}{PAC} = \overset{\frown}{XC}$.

NOTE: By similar reasoning, CH is perpendicular to AB, and thus all three altitudes of $\triangle ABC$ are concurrent at H. Observe that the reason H lies above

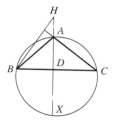

Figure 2.6

A on line AD in Figure 2.6 is that $\angle A$ is obtuse. This exercise, together with Problem 2.6, shows that in all cases, the altitudes of a triangle are concurrent.

2A.5 In the situation of Figure 2.6, show that the reflection of H in line AC lies on the circle.

HINT: Let P be as in the hint for the previous exercise and show that P is the reflection of H in AC by proving that $\triangle HAP$ is isosceles. Observe that $\angle C$ and $\angle HPA$ are both supplementary to $\angle APB$.

NOTE: We now see that even when $\triangle ABC$ is not acute angled, the three reflections of the orthocenter H in the sides of the triangle all lie on the circumcircle.

2B   The Centroid

In every triangle, the three medians are concurrent. This fact appeared as Exercise 1H.2, but we present a better proof of a stronger assertion as Theorem 2.7. The point of concurrence of the medians is called the **centroid** of the triangle, and it is customarily denoted G.

(2.7) THEOREM. *The three medians of an arbitrary triangle are concurrent at a point that lies two thirds of the way along each median from the vertex of the triangle toward the midpoint of the opposite side.*

Proof.   Let G be the point where median AX crosses median BY, as in Figure 2.7. We first show that G lies two thirds of the way from A to X along line segment AX. We need to show, in other words, that $AG = 2GX$.

Draw XY and observe that this segment joins the midpoints of two sides of $\triangle ABC$. By Corollary 1.31, we conclude that XY must be parallel to the third side, AB, and that $XY = \frac{1}{2}AB$. Since $XY \parallel AB$, we have equality of alternate interior angles, and thus $\angle BAG = \angle YXG$ and $\angle ABG = \angle XYG$. It follows that $\triangle BAG \sim \triangle YXG$ by AA, and hence $AG/GX = AB/XY = 2$, as required.

Similarly, median CZ crosses AX at a point that lies two thirds of the way from A to X. Clearly, however, G is the only point on AX that has this property, and thus CZ goes through G. The three medians are thus concurrent, and we know that the point of concurrence lies two thirds of the way along median AX. By similar reasoning, we deduce that the point of concurrence lies two thirds of the way along each median. ∎

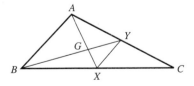

Figure 2.7

As an application, we have the following converse to the easy observation that the medians to the two equal sides of an isosceles triangle are equal. (See Exercise 1B.4.)

(2.8) PROBLEM.   Suppose that in $\triangle ABC$, medians BY and CZ have equal lengths. Prove that $AB = AC$.

Solution.   Medians BY and CZ intersect at the centroid G, and we have

$$BG = \frac{2}{3}BY = \frac{2}{3}CZ = CG$$

by Theorem 2.7. Thus $\triangle BGC$ is isosceles and we have $\angle GBC = \angle GCB$ by the pons asinorum.

We are given that $BY = CZ$, and of course, $BC = BC$. Since we now know that $\angle YBC = \angle ZCB$, it follows that $\triangle BYC \cong \triangle CZB$ by SAS, and thus $YC = ZB$. We conclude that $AB = 2ZB = 2YC = AC$, as required.   ∎

By Problem 2.8, we know that a triangle having two equal medians must be isosceles, and in Exercise 1B.7, we saw that a triangle having two equal altitudes is isosceles. (This was proved more fully using area principles in Section 1E.) These two results suggest the possibility that the equality of two angle bisectors of a triangle also guarantees that the triangle is isosceles. This is true but notoriously difficult to establish. A proof appears in Section 2D.

The centroid of a triangle has a significance that can be explained by a little physics experiment, which we encourage readers to try. Cut out a triangle from a uniform piece of cardboard and then use one thumbtack to attach the triangle to a wall or to a bulletin board in such a way that the triangle can swing freely in a vertical plane. To overcome friction, it may be necessary to use a somewhat longer tack so that the cardboard does not rub against the wall. The interesting fact is that the cardboard triangle will come to rest with its centroid directly below the point of suspension, and this is true regardless of where we choose the point of suspension to be, although in practice, this works best when the suspension point is relatively far from the centroid.

To carry out the experiment, determine the centroid by carefully drawing two medians and clearly marking their intersection point. To help see that the centroid comes to rest directly below the point of suspension, it is useful to tie a thread to the tack in front of the cardboard triangle. If a weight is attached to the free end of the thread, the thread will hang vertically, and so it should pass directly in front of the marked centroid of the triangle.

Why does this work? What we are really saying here is that the center of mass of a uniform triangular sheet is located at the centroid of the triangle. We expect that most readers of this book will know about centers of mass from their study of calculus or physics. In particular, readers may remember that every rigid physical body behaves in static balancing experiments as though all of its mass were concentrated at a single point called the **center of mass** of the body. To prove that the center of mass M of a uniform cardboard triangle coincides with the centroid, it suffices to show that M lies on each median.

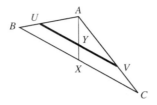

Figure 2.8

To see that the center of mass of the region bounded by $\triangle ABC$ lies on median AX, for example, imagine slicing the triangular region into a huge number of very thin strips, each parallel to BC. In Figure 2.8, line segment UV represents one of these strips. Observe that the point Y, where strip UV meets median AX, is actually the midpoint of UV. This fact, proved using similar triangles, is contained in Exercise 1H.8 and the note following it. Clearly, the center of mass of strip UV is at its midpoint Y, and thus the strip behaves as though all of its mass were concentrated at Y. Since Y lies on median AX, we see that each of the pieces into which our triangle has been decomposed appears to have its mass concentrated on this median. We can pretend, therefore, that the entire mass of the triangle is distributed along AX, and it follows that the center of mass M lies on this median. Similarly, of course, M also lies on the other two medians, and thus it coincides with the centroid, as claimed.

Because of its appeal to physical intuition, the foregoing argument should not, of course, be considered a rigorous mathematical proof. It is interesting to note, however, that this physical reasoning demonstrates that there is a particular point (the center of mass) that lies on each of the medians and thus provides an alternative argument to show that the medians of a triangle must be concurrent.

We shall see other examples where physical reasoning can be used to "prove" geometry theorems. There is a simple physics argument, for example, that demonstrates that the centroid must lie two thirds of the way along each median. This time, imagine that our $\triangle ABC$ is made from some massless rigid material and that a unit point mass is attached at each vertex. There are two equal masses at the ends of side BC, and so that side behaves as though its mass were concentrated at the midpoint X. We can thus pretend that the three units of mass of the entire triangle are positioned with one unit at A and two units at X. The center of mass M of the triangle, therefore, lies on AX, closer to X than to A. More precisely, we see that since the mass at X is twice that at A, the distance $MX = \frac{1}{2}MA$. In other words, M lies on AX, two thirds of the way from A to X. Since M also lies on the other medians by similar reasoning, it follows that M is the centroid and that it lies two thirds of the way along each median.

We have seen that the center of mass of a uniform cardboard triangle is at the centroid and also that the center of mass of a massless triangle with equal point masses at the vertices is at the centroid. Another interesting case is that of a triangle made of uniform wire. We assume, in other words, that the mass is uniformly distributed along the sides of $\triangle ABC$ and that the interior is massless. In this case, we can assume that the total mass of side BC is a (the length of BC), and we pretend that it is concentrated at the midpoint X of BC. Similarly, a mass of b units is at the midpoint Y of AC, and a mass of c units is at the midpoint Z of AB. We now need to consider only $\triangle XYZ$ with point masses a, b, and c at vertices X, Y, and Z. We mention that $\triangle XYZ$, the triangle formed by the midpoints of the sides of $\triangle ABC$, is called the **medial triangle** of $\triangle ABC$. By Corollary 1.31, we see that $XY = c/2$, $XZ = b/2$, and $YZ = a/2$, and thus $\triangle ABC \sim \triangle XYZ$ by SSS.

We can now replace the two point masses at Y and Z by a single mass at the center of mass P of side YZ. Since the masses at Y and Z are b and c, and these are not necessarily equal, point P need not be the midpoint of YZ. To locate P precisely, let $YP = u$ and $ZP = v$. Then $bu = cv$, and thus $u/v = c/b = XY/XZ$. Hence we see that point P divides side YZ of $\triangle XYZ$ into two pieces whose lengths are in the same ratio as the lengths of the nearer sides of the triangle. Comparison of this with Theorem 1.12 shows that the bisector of $\angle X$ meets YZ at P, and thus the center of mass of $\triangle XYZ$, with the appropriate point masses at the vertices, lies on the angle bisector XP. Similarly, this center of mass also lies on the other two angle bisectors of $\triangle XYZ$. It follows that the center of mass of our original uniform wire triangle lies at the point of concurrence of the angle bisectors of the medial triangle. We shall see in Section 2E that the angle bisectors of an arbitrary triangle are concurrent at a point called the **incenter**. In general, it is not true that the incenter of the medial triangle is the centroid of the original triangle, and thus the center of mass of a uniform wire triangle is not always at the same location as the center of mass of the corresponding uniform cardboard triangle.

Exercises 2B

2B.1 Show that the centroid of the medial triangle of $\triangle ABC$ is the centroid of $\triangle ABC$.

2B.2 Show using centers of mass that the angle bisectors of $\triangle ABC$ are concurrent at a point I, where I lies on bisector AP at a position such that

$$\frac{AI}{AP} = \frac{b+c}{a+b+c},$$

where, as usual, a, b, and c denote the lengths of the sides of the triangle.

HINT: Put point masses of a, b, and c units at points A, B, and C.

2B.3 Show that there can be no point P in the interior of $\triangle ABC$ such that every line through P subdivides the triangle into two pieces of equal area.

HINT: If there is a point P that has this property, show that the medians would have to go through P.

NOTE: The fact that the center of mass of a uniform triangular sheet lies at the centroid certainly does not imply that every line through this point divides the area into two equal parts.

2B.4 Suppose that a convex quadrilateral $ABCD$ is cut from a uniform piece of cardboard. Show that the center of mass of the cardboard quadrilateral lies on diagonal AC if and only if AC bisects diagonal BD.

HINT: Consider the line joining the centers of mass of $\triangle ABD$ and $\triangle CBD$.

2B.5 Choose a point P on side AC of $\triangle ABC$ and write $p = AP/AC$. Let $x = AX/AM$, where X is the point where BP crosses median AM, and observe that $x = 2/3$ when $p = 1/2$. Find a general formula for x in terms of p.

HINT: Put masses of one unit each at points B and C and find an appropriate mass to put at point A so that the center of mass will be at X.

2C   The Euler Line, Orthocenter, and Nine-Point Circle

So far, we have discussed two important points associated with a triangle: the circumcenter and the centroid. If the given triangle is equilateral, then each median is the perpendicular bisector of the opposite side, and it follows that the circumcenter and the centroid are actually the same point. In all other cases, however, these points are distinct.

(2.9) LEMMA.   *If the circumcenter and the centroid of a triangle coincide, then the triangle must be equilateral.*

Proof.   Suppose that the centroid of $\triangle ABC$ is also the circumcenter. Let G be the centroid and let X be the midpoint of side BC. Then X, G, and A are distinct points on the median from A, and since these points are collinear, we see that A lies on line GX. But G is also the circumcenter, and so it lies on the perpendicular bisector of side BC. Of course, the midpoint X of side BC also lies on the perpendicular bisector of BC, and it follows that the line GX is the perpendicular bisector of BC. Since A lies on GX, we have shown that A lies on the perpendicular bisector of BC, and thus A is equidistant from B and C. We now have $AB = AC$, and similar reasoning shows that $BA = BC$. It follows that all three sides of $\triangle ABC$ are equal, and thus the triangle is equilateral.   ∎

Given any nonequilateral $\triangle ABC$, we now know that the circumcenter O and the centroid G are two distinct points. These two points determine a unique line that is called the **Euler line** of the triangle. (Leonhard Euler was a Swiss astronomer and mathematician who lived from 1707 to 1783. His name is pronounced "oiler.") We stress that for equilateral triangles, there is no Euler line defined.

We have already mentioned that the three altitudes of a triangle are always concurrent at a point called the **orthocenter** of the triangle. Remarkably, the Euler line also goes through the orthocenter. In some sense, it is a double miracle when four previously determined lines are concurrent.

For an equilateral triangle, the altitudes are the medians, and we know that they are concurrent at the centroid. Since there is no Euler line for an equilateral triangle, we have nothing further to prove in this case. For nonequilateral triangles, we want to show that all three altitudes meet at some point of the Euler line. In fact, we are able to specify this point in advance.

(2.10) THEOREM.   *Assume that* $\triangle ABC$ *is not equilateral and let G and O be its centroid and circumcenter, respectively. Let H be the point on the Euler line GO that lies on the opposite side of G from O and such that* $HG = 2GO$. *Then all three altitudes of* $\triangle ABC$ *pass through H.*

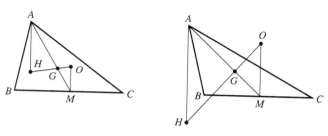

Figure 2.9

Proof.   We shall show that the altitude from A passes through H; the proofs for the other two altitudes are similar, of course. If H coincides with A, there is nothing to prove, and so we can assume that H and A are different points. But note that it really can happen that H and A coincide; this occurs when $\angle A = 90°$. It suffices now to show that line AH is perpendicular to BC.

Let M be the midpoint of side BC and observe that O and M are distinct points. Otherwise, the median GM is the Euler line, and since we know that $AG = 2GM$, it would follow that A is H, which we are assuming is not the case. Since both O and M lie on the perpendicular bisector of BC, line OM is perpendicular to BC, and we will be done if we can show that OM is parallel to AH.

Figure 2.9 shows two of the several possible configurations for this problem, but the proof is identical in all cases. We will establish that $AH \parallel OM$ by proving the equality of the alternate interior angles, $\angle H$ and $\angle O$. We know that $AG/MG = 2$ by Theorem 2.7 and that $HG/OG = 2$ by the construction of the point H. Since $AG/MG = HG/OG$ and $\angle AGH = \angle MGO$, we see that $\triangle AGH \sim \triangle MGO$ by SAS. It follows that $\angle H = \angle O$, and the proof is complete. ∎

Given $\triangle ABC$, let H be its orthocenter. If we start with a right triangle, then clearly H coincides with the right angle, and in all other cases, A, B, C, and H are easily seen to be four distinct points. Observe that the line determined by any two of these four points is perpendicular to the line determined by the other two. For example, AH is perpendicular to BC because line AH is the altitude from A in $\triangle ABC$. It is amusing to observe that each of the four points A, B, C, and H is the orthocenter of the triangle formed by the other three. That A is the orthocenter of $\triangle HBC$, for example, is just another way of saying that AH, AB, and AC are perpendicular to BC, HC, and HB, respectively, which is a fact that we have already noted.

Given any four points with the property that each is the orthocenter of the triangle formed by the other three, we say that the given set of four points is an **orthic quadruple**. We now know that given an arbitrary set of three points A, B, and C, there almost always exists a fourth point H such that the set $\{A, B, C, H\}$ is an orthic quadruple. (Just take H

to be the orthocenter of $\triangle ABC$.) The only exceptions are when $\triangle ABC$ is a right triangle or when $\triangle ABC$ does not exist because the given three points are collinear.

Given $\triangle ABC$, let D, E, and F be the points where the altitudes from A, B, and C meet lines BC, AC, and AB, respectively. These points are called the **feet** of the altitudes, and we note that they may not actually lie on the line *segments* BC, AC, and AB. If $\triangle ABC$ is not a right triangle, it is not hard to see that the feet D, E, and F are distinct and form a triangle. We refer to $\triangle DEF$ as the **pedal triangle** of $\triangle ABC$, although some authors call it the **orthic triangle**.

In Figure 2.10, we have drawn two of the situations that can occur. In each diagram, the original triangle is drawn with heavy lines. In the configuration on the right, we had to extend two of the sides of the triangle to meet the altitudes, and these extensions are shown with dashes. In both diagrams, the pedal triangle and the altitudes are drawn with solid lighter lines. It should be clear from the figure that the feet of the altitudes lie on the sides of the triangle for acute angled triangles, but two of the feet lie outside of the triangle if there is an obtuse angle.

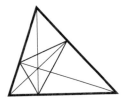 

Figure 2.10

Figure 2.10 shows something else of interest. The triangle on the left and the one on the right happen to be two of the four triangles that can be formed using three of the four points of an orthic quadruple. We see that in both cases, we get exactly the same pedal triangle.

(2.11) THEOREM.   *The pedal triangles of each of the four triangles determined by an orthic quadruple are all the same.*

Proof.   We have seen previously that for each choice of two of our four given points, the line determined by those two is perpendicular to the line determined by the other two. Since there are exactly three ways to pair off four objects into two sets of two, this gives three points that occur as the intersections of pairs of perpendicular lines determined by our orthic quadruple. It is easy to see that these three points must be the vertices of the pedal triangle of each of the four triangles.   ∎

The circumcircle of the pedal triangle of $\triangle ABC$ turns out to be an amazing object. In addition to the feet of the three altitudes, this circle also contains the midpoints of the three sides; hence it is also the circumcircle of the medial triangle of $\triangle ABC$. Remarkably, this circle has a further unexpected property: It bisects each of the line segments AH, BH, and CH, where H is the orthocenter of $\triangle ABC$. We refer to the midpoint X of segment AH as the **Euler point** of $\triangle ABC$ opposite to side BC, and similarly, the midpoints Y and Z of BH and CH are the Euler points of $\triangle ABC$ opposite to sides AC and AB, respectively. The common circumcircle of the pedal and medial triangles contains the three Euler points and even more is true. We can state the full result as follows.

(2.12) THEOREM. *Given any triangle, all of the following points lie on a common circle: the three feet of the altitudes, the three midpoints of the sides, and the three Euler points. Furthermore, each of the line segments joining an Euler point to the midpoint of the opposite side is a diameter of this circle.*

The remarkable circle whose existence is asserted in Theorem 2.12 is called the **nine-point circle** of the triangle. We should point out, however, that the nine points referred to in the statement of the theorem are not always distinct.

Proof of Theorem 2.12.   In Figure 2.11, we have drawn one possible configuration for this theorem. The proof will be the same for all cases, although the diagram can look a little different for different starting triangles. In the figure, points D, E, and F are the feet of the altitudes of $\triangle ABC$. Points P, Q, and R are the midpoints of the sides, and X, Y, and Z are the Euler points. We need to show that all nine of these points lie on a common circle and that XP, YQ, and ZR are diameters of this circle. As usual, we have called the orthocenter H.

Draw line segment YQ and consider the unique circle that has YQ as a diameter. We will first show that points P and R lie on this circle. To see that P lies on the circle with diameter YQ, it suffices to show that $\angle YPQ = 90°$. We will do this by proving that $YP \parallel CF$ and $PQ \parallel AB$. Since CF and AB are perpendicular, it follows easily that YP and PQ are perpendicular, and thus $\angle YPQ = 90°$, as required.

That PQ is parallel to AB is clear by Corollary 1.31 since P and Q are the midpoints of two sides of $\triangle ABC$ and AB is the third side. Similarly, to prove that YP is parallel to CF, we work in $\triangle BHC$. Of course, P is the midpoint of side BC of this triangle, and the Euler point Y is the midpoint of side BH. It follows that YP is parallel to the third side of this triangle, which is CH, and thus $YP \parallel CF$, as desired. We have now shown that $\angle YPQ = 90°$, and thus point P lies on the circle with diameter YQ, as desired.

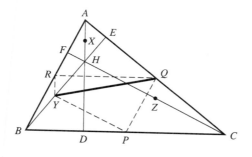

Figure 2.11

Next, we use similar reasoning to show that R lies on this circle. It suffices to prove that $\angle YRQ = 90°$, and we accomplish this by showing that $YR \| AD$ and $QR \| BC$. Since altitude AD is perpendicular to side BC, it will follow that QR is perpendicular to YR, and thus $\angle YRQ = 90°$, as claimed. That $QR \| BC$ is a straightforward application of Corollary 1.31 in $\triangle ABC$, and to prove that $YR \| AD$, we apply Corollary 1.31 in $\triangle ABH$.

We have now shown that points P, Q, R, and Y all lie on the same circle and that YQ is a diameter of this circle. Since this circle contains P, Q, and R, it is, of course, the circumcircle of the medial triangle of $\triangle ABC$. We have now proved that given an arbitrary $\triangle ABC$, the circumcircle of its medial triangle has line segment QY as a diameter, where Q is the midpoint of side AC and Y is the opposite Euler point. It follows similarly that the line segments PX and RZ are also diameters of the medial circumcircle, and in particular, the other two Euler points, X and Z, lie on this circle. We have now shown that six of the required nine points lie on the medial circumcircle and that each of XP, YQ, and ZR is a diameter of this circle.

Next, we observe that E lies on the circle with diameter YQ since $\angle YEQ = 90°$. This shows that the altitude foot E lies on the medial circumcircle of an arbitrary $\triangle ABC$. It follows similarly that the medial circumcircle contains the other two altitude feet, D and F, and this concludes the proof. ∎

(2.13) PROBLEM.   What is the radius and where is the center of the nine-point circle of $\triangle ABC$?

Perhaps the easiest way to analyze this situation is via the technique of transformational geometry, which we introduce informally as we proceed.

Solution to Problem 2.13.   Since the nine-point circle is, among other things, the circumcircle of the medial triangle of $\triangle ABC$, we can solve this problem by focusing attention on $\triangle PQR$, where P, Q, and R are the midpoints of sides BC, AC, and AB, as in Figure 2.12.

We know that $QP \| AR$ and $RP \| AQ$; thus $AQPR$ is a parallelogram, and diagonal AP bisects diagonal RQ by Theorem 1.9. In other words, median AP of $\triangle ABC$ bisects side RQ of the medial triangle PQR, and hence it contains the median from P in triangle PQR. Similarly, the other two medians of $\triangle ABC$

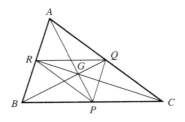

Figure 2.12

contain the other medians of $\triangle PQR$, and it follows that the centroid G of $\triangle ABC$ is also the centroid of $\triangle PQR$.

Now imagine the following two-step transformation of the plane of Figure 2.12. First, shrink the plane with a scale factor $\lambda = .5$ in such a way that point G remains fixed and every other point moves toward G. In other words, an arbitrary point X of the plane is moved to the midpoint of segment GX. Next, rotate the plane 180° with a center of rotation at G. It does not matter, of course, if this rotation is clockwise or counterclockwise. Write $\mathbf{T}$ to denote the net effect of these two operations and view $\mathbf{T}$ as a function. Observe that $\mathbf{T}(A) = P$ since $AG = 2GP$ and $\angle AGP = 180°$. Similarly, $\mathbf{T}(B) = Q$ and $\mathbf{T}(C) = R$.

We now argue without a formal proof that the transformation $\mathbf{T}$ carries lines to lines, triangles to triangles, and circles to circles. Furthermore, given a circle centered at some point O, the image of that circle under $\mathbf{T}$ is a circle centered at the point $\mathbf{T}(O)$ and having radius half that of the original circle. Of course, all of this can be formalized and made more precise, and rigorous proofs can be constructed. We trust, however, that the truth of our assertions is reasonably clear, and we proceed without further proof.

Now consider the circumcircle of $\triangle ABC$, which is centered at the circumcenter O and has radius R. The transformation $\mathbf{T}$ carries $\triangle ABC$ to $\triangle PQR$, and hence it carries the circumcircle of $\triangle ABC$ to the circumcircle of $\triangle PQR$, which is the nine-point circle of $\triangle ABC$. It follows that the center of the nine-point circle, which we call N, is exactly the point $\mathbf{T}(O)$ and the radius of the nine-point circle is $R/2$. Also, since XP, YQ, and ZR are diameters of the nine-point circle, it follows that each of these segments has length R. We record this information here for future use. ∎

(2.14) THEOREM.   *Let R be the circumradius of $\triangle ABC$. Then the distance from each Euler point of $\triangle ABC$ to the midpoint of the opposite side is R, and the radius of the nine-point circle of $\triangle ABC$ is $R/2$.* ∎

(2.15) COROLLARY.   *Suppose $\triangle ABC$ is not a right triangle and let H be its orthocenter. Then $\triangle ABC$, $\triangle HBC$, $\triangle AHC$, and $\triangle ABH$ have equal circumradii.*

Proof.   We know by Theorem 2.11 that these four triangles share a common pedal triangle. Since the nine-point circle of any triangle is the circumcircle of its pedal triangle, it follows that the four triangles share a common nine-point circle. We have just seen, however, that for an arbitrary triangle, the circumradius is exactly twice the nine-point radius, and it follows that our four triangles have equal circumradii. ∎

Solution to Problem 2.13, continued.   Our goal now is to give a convenient description of the location of the nine-point center N of $\triangle ABC$. We know, of course, that N must be the midpoint of each of the segments PX, QY, and RZ, but we are looking for a more useful characterization. In the case where the given triangle is equilateral, it is clear that N coincides with G and O and H. We assume, therefore, that the given triangle is not equilateral, and thus it has an Euler line GO. We know that

$N = \mathbf{T}(O)$, and it follows from the definition of $\mathbf{T}$ that N lies on the line through G and O, which is the Euler line. Furthermore, N lies on the opposite side of G from O, and we have $NG = \frac{1}{2}GO$. Recall now that the orthocenter H also lies on the Euler line on the opposite side of G from O and that $HG = 2GO$. It follows that N lies on the segment GH and $HN = HG - NG = \frac{3}{2}GO$. Also, $NO = NG + GO = \frac{3}{2}GO$, and we deduce that $HN = NO$. In other words, the nine-point center N is the midpoint of the segment HO. Note that this is true, in some sense, even if $\triangle ABC$ is equilateral since in that case N, H, and O are all the same point. This completes our solution to Problem 2.13.   ∎

(2.16) COROLLARY. *Suppose $\triangle ABC$ is not a right triangle and let H be its orthocenter. Then the Euler lines of $\triangle ABC$, $\triangle HBC$, $\triangle AHC$, and $\triangle ABH$ are concurrent. If any of these triangles is equilateral, then the Euler lines of the remaining triangles are concurrent.*

Proof.   We have seen that these four triangles share a nine-point circle. The center N of this circle lies on all of the Euler lines, which are therefore concurrent.   ∎

Exercises 2C

2C.1 Suppose that the Euler line of $\triangle ABC$ is perpendicular to BC. Show that $AB = AC$.

2C.2 Show that the circumcenter of $\triangle ABC$ is the orthocenter of the medial triangle.

2C.3 Given $\triangle ABC$, draw line WV through A parallel to BC, line UW through B parallel to AC, and line UV through C parallel to AB. Show that the orthocenter of $\triangle ABC$ is the circumcenter of $\triangle UVW$.

2C.4 Show that the nine-point circle of $\triangle ABC$ is the locus of all midpoints of segments UH, where H is the orthocenter of $\triangle ABC$ and U is an arbitrary point of the circumcircle.

HINT: By Exercise 1H.10, we already know that this locus is a circle.

2C.5 Given $\triangle ABC$, let X be the Euler point opposite side BC. Show that length AX is equal to the perpendicular distance from the circumcenter O of $\triangle ABC$ to side BC. Deduce that $(AH)^2 = 4R^2 - a^2$, where as usual, H is the orthocenter, R is the circumradius, and $a = BC$.

HINT: Show that the diagonals of quadrilateral $XHPO$ bisect each other, where P is the midpoint of side BC.

NOTE: Imagine holding points B and C fixed and letting point A move on some fixed circle through B and C. Then H is a variable point, but this problem shows that the length of the variable line segment AH is constant, independent of the choice of A. Note also that AH will always be perpendicular to BC.

2C.6 Given four points on a circle, join each of them to the orthocenter of the triangle formed by the other three. Show that the four line segments that result have a common midpoint M, and so in particular, these four lines are concurrent at M. Show also that the nine-point circles of the four triangles formed by each three of the four points all go through M.

HINT: Consider just two of the line segments and show that they are the diagonals of a parallelogram. Use Exercise 2C.4 for the second part.

2D   Computations

Suppose we are told that the lengths of the sides of $\triangle ABC$ are the three numbers a, b, and c. Without being given any additional data and without even being allowed to see the triangle, we are asked to determine its area. In theory, at least, we can see that this should be possible because we could draw our own triangle having the same three side lengths, and then we could measure its area. Since our triangle is congruent to the original one by SSS, it follows that the area we measured is equal to the area of the unseen $\triangle ABC$. Similarly, it is in principle possible to determine the circumradius of $\triangle ABC$, or to determine any other numerical quantity associated with the triangle, when the only data we have available are the lengths of the sides.

In this section, we will show how to compute the area of a triangle given the lengths of its sides. More generally, our goal is to develop formulas relating various numerical quantities associated with a triangle. The law of sines is an example of the sort of thing we want; it relates the side lengths and the angles of a triangle, and in its extended form, it relates these to the circumradius. Another example, proved in Corollary 2.5, is the equation $4KR = abc$ relating the area, circumradius, and side lengths of a triangle.

We begin with the following question: Given the side lengths of $\triangle ABC$, how can the angles be determined? Perhaps the easiest way is via the law of cosines.

(2.17) THEOREM (Law of Cosines).   *Given $\triangle ABC$, let a, b, and c denote, as usual, the lengths of sides BC, AC, and AB, respectively. Then*

$$c^2 = a^2 + b^2 - 2ab\cos(C).$$

Given a, b, and c, the equation of the law of cosines can easily be solved to obtain $\cos(C) = (a^2 + b^2 - c^2)/2ab$, and of course, similar formulas yield $\cos(A)$ and $\cos(B)$ in terms of a, b, and c.

Instead of proceeding immediately to give a proof of Theorem 2.17, we have decided to devote a few sentences to a discussion of how one might find a proof. It is often useful to think about special cases to help try to understand what is really going on in some formula. Consider, for example, what happens in the law of cosines if $\angle C = 90°$. Then $\cos(C) = 0$, and in this situation, the formula tells us that $c^2 = a^2 + b^2$, which we recognize as the Pythagorean theorem. The law of cosines thus includes the Pythagorean theorem as a special case, and it follows that we should try to use the Pythagorean theorem in the proof of Theorem 2.17. Otherwise, we will find ourselves reproving this theorem within the proof of the law of cosines. To use the Pythagorean theorem, however, we need a right triangle, and so it would seem that a good first step is to draw an altitude of our given triangle.

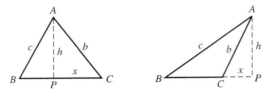

Figure 2.13

Proof of Theorem 2.17.   Since a triangle can have at most one angle that fails to be acute, we can be sure that at least one of $\angle A$ or $\angle B$ is an acute angle, and it is therefore no loss to assume that $\angle B < 90°$. Draw altitude AP from A to BC and write $h = AP$. Note that there are three possibilities: Either $\angle C < 90°$ and P lies on segment BC, as in the left diagram of Figure 2.13, or $\angle C = 90°$ and point P coincides with point C, or $\angle C > 90°$ and P lies on an extension of side BC, as in the right diagram of Figure 2.13.

Suppose first that $\angle C < 90°$ and write $x = PC$. Since $\triangle APC$ is a right triangle, we see that $\cos(C) = PC/AC = x/b$, and so $x = b\cos(C)$. Also, two applications of the Pythagorean theorem yield $b^2 = x^2 + h^2$ and $c^2 = h^2 + (a-x)^2$. This gives

$$c^2 = (b^2 - x^2) + (a - x)^2 = b^2 - x^2 + a^2 - 2ax + x^2 = a^2 + b^2 - 2ax .$$

Since $x = b\cos(C)$, we obtain the desired formula.

We have already seen that the law of cosines holds when $\angle C = 90°$, and so we can now assume that $\angle C > 90°$, and we refer to the diagram on the right of Figure 2.13. Here too we write $x = PC$, but in this case, we have $\cos(C) = -x/b$ and $x = -b\cos(C)$. Two applications of the Pythagorean theorem yield $b^2 = x^2 + h^2$ and $c^2 = h^2 + (a + x)^2$, and we have

$$c^2 = (b^2 - x^2) + (a + x)^2 = b^2 - x^2 + a^2 + 2ax + x^2 = a^2 + b^2 + 2ax .$$

Since $x = -b\cos(C)$ in this case, the proof is complete.   ∎

As an application of the law of cosines, we can derive Heron's formula for the area of a triangle given the lengths of the sides. To state the formula cleanly, we introduce a new quantity $s = \frac{1}{2}(a + b + c)$ called the **semiperimeter**.

(2.18) THEOREM (Heron's Formula).   *The area K of $\triangle ABC$ is given by the equation*

$$K = \sqrt{s(s - a)(s - b)(s - c)},$$

where a, b, and c are the lengths of the sides and s is the semiperimeter.

It may be of some value to try out Heron's formula in a couple of cases where we already know how to compute the area. Suppose, for example, that $\triangle ABC$ is a right triangle with arms of length 3 and 4 and hypotenuse of length 5. We know that $K = \frac{1}{2}bh = \frac{1}{2}(3)(4) = 6$ in this case. On the other hand, we have $s = \frac{1}{2}(3 + 4 + 5) = 6$, and Heron's formula gives $K = \sqrt{(6)(6-3)(6-4)(6-5)} = \sqrt{36} = 6$, as expected. Next, consider an equilateral triangle, each of whose sides has length 2. It is easy to see that $K = \frac{1}{2}bh = \frac{1}{2}(2)(\sqrt{3}) = \sqrt{3}$, and Heron's formula gives $K = \sqrt{(3)(3-2)(3-2)(3-2)} = \sqrt{3}$, again as expected.

Proof of Theorem 2.18.   We know that $K = \frac{1}{2}ab\sin(C)$, and so

$$4K^2 = a^2b^2\sin^2(C) = a^2b^2(1 - \cos^2(C)).$$

The law of cosines gives $\cos(C) = (c^2 - a^2 - b^2)/2ab$, and we can substitute this into the previous equation to obtain

$$4K^2 = a^2b^2\left(1 - \frac{(c^2 - a^2 - b^2)^2}{4a^2b^2}\right) = a^2b^2 - \frac{(c^2 - a^2 - b^2)^2}{4}.$$

Thus $16K^2 = 4a^2b^2 - (c^2 - a^2 - b^2)^2$, and the right side of this equation factors as a difference of squares to yield

$$16K^2 = [2ab + (c^2 - a^2 - b^2)][(2ab - (c^2 - a^2 - b^2)]$$
$$= [(c^2 - (a-b)^2][(a+b)^2 - c^2].$$

Each of the factors on the right of the previous equation factors as a difference of squares, and we obtain

$$16K^2 = [c + (a - b)][(c - (a - b)][(a + b) + c][(a + b) - c].$$

Observe that $c + a - b = (a + b + c) - 2b = 2(s - b)$. Similarly, the second factor in our formula for $16K^2$ equals $2(s - a)$, and the third and fourth factors are $2s$ and $2(s - c)$, respectively. It follows that $K^2 = s(s - a)(s - b)(s - c)$, and Heron's formula follows.   ∎

(2.19) PROBLEM.   Express the circumradius R of $\triangle ABC$ in terms of the lengths of the sides.

Solution.   This is easy. We know that $4KR = abc$, and hence

$$R = \frac{abc}{4K} = \frac{abc}{4\sqrt{s(s-a)(s-b)(s-c)}}.   ∎$$

Next, we offer another application of the law of cosines. In Figure 2.14, point P was chosen arbitrarily on side AB of $\triangle ABC$, dividing the side, which is of length c, into pieces with lengths x and y, as shown. The following result provides a formula that allows one to compute the length $t = CP$ in terms of the given data: x, y, $b = AC$, and $a = BC$. Of course, we also have $c = AB = x + y$.

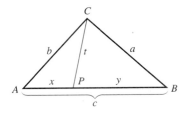

Figure 2.14

(2.20) THEOREM (Stewart). *In the situation of Figure 2.14, we have*

$$ct^2 + xyc = xa^2 + yb^2.$$

Proof. By the law of cosines, we have

$$t^2 = a^2 + y^2 - 2ay\cos(B) \quad \text{and}$$
$$b^2 = a^2 + c^2 - 2ac\cos(B).$$

We can eliminate $\cos(B)$ if we multiply the first equation by c and the second by y and then subtract. Using the fact that $c - y = x$, we obtain

$$ct^2 - yb^2 = (c - y)a^2 + cy^2 - yc^2 = xa^2 + cy(y - c) = xa^2 - xyc,$$

and the desired equation follows. ∎

(2.21) PROBLEM. Show how to compute the lengths of the angle bisectors of a triangle given the lengths of the sides.

Solution. To use Stewart's theorem to find the length t of angle bisector CP of $\triangle ABC$ in terms of a, b, and c, we need to express x and y in terms of the given data. We recall from Theorem 1.12 that bisector CP divides AB into pieces proportional to the lengths of the nearer sides. We have, in other words, $x/y = b/a$. Since $x + y = c$, a bit of algebra yields $x = bc/(a + b)$ and $y = ac/(a + b)$. (To check the algebra, observe that the sum of these two fractions is c and that the first divided by the second is equal to b/a.) It follows that

$$xyc = \frac{abc^3}{(a+b)^2} \quad \text{and} \quad xa^2 + yb^2 = \frac{a^2bc + b^2ac}{a+b} = abc$$

and we can substitute into the Stewart's theorem equation to get

$$ct^2 + \frac{abc^3}{(a+b)^2} = abc.$$

A little more algebra now yields

$$t^2 = ab\left[1 - \frac{c^2}{(a+b)^2}\right],$$

and the length t of the angle bisector can be found by taking the square root. ∎

As a check of this rather unpleasant formula, we can try it in the case of an isosceles triangle, where $a = b$ and the angle bisector CP is an altitude. The Pythagorean theorem yields $t^2 = a^2 - c^2/4$, and a bit of algebra shows that this agrees with our previous calculation.

(2.22) PROBLEM. Let AX and BY be angle bisectors in $\triangle ABC$ and suppose that $AX = BY$. Show that $AC = BC$.

We saw in Exercise 1B.7 that if two altitudes of a triangle are equal, then the triangle is isosceles, and in Problem 2.8, we proved the corresponding assertion for medians. Problem 2.22, which is the analog of these facts for angle bisectors, seems by far the hardest of these three similar-sounding results. One unusual aspect of the following argument is that we prove an equality by contradiction.

Solution to Problem 2.22. Changing the roles of a, b, and c appropriately, we can apply the formula derived in Problem 2.21 to compute that

$$(AX)^2 = bc \left[1 - \frac{a^2}{(b+c)^2} \right] \quad \text{and}$$

$$(BY)^2 = ac \left[1 - \frac{b^2}{(a+c)^2} \right],$$

and by hypothesis, we know that these two quantities are equal. Division by c thus yields

$$b \left[1 - \frac{a^2}{(b+c)^2} \right] = a \left[1 - \frac{b^2}{(a+c)^2} \right],$$

and we get

$$b - a = \frac{ba^2}{(b+c)^2} - \frac{ab^2}{(a+c)^2}.$$

We need to show that $a = b$, and so we try to derive a contradiction by supposing that a and b are unequal. It is no loss to assume that $b > a$, and so the left side of the previous equation is positive. The right side must therefore also be positive, and it follows that

$$\frac{a}{(b+c)^2} > \frac{b}{(a+c)^2}.$$

But $b > a$, and so

$$\frac{b}{(a+c)^2} > \frac{a}{(a+c)^2} > \frac{a}{(b+c)^2}.$$

This contradicts the previous inequality and thereby proves the result. ∎

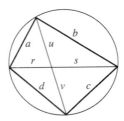

Figure 2.15

We close this section by using Stewart's theorem to deduce one more computational result.

(2.23) THEOREM (Ptolemy).   *Let a, b, c, and d be the lengths of consecutive sides of a quadrilateral inscribed in a circle and suppose x and y are the lengths of the diagonals. Then $ac + bd = xy$.*

Proof.   Let r, s, u, and v be the lengths of the four partial diagonals, as shown in Figure 2.15, where $r + s = x$ and $u + v = y$. It is easy to see from the two pairs of similar triangles that

$$\frac{a}{c} = \frac{u}{s} = \frac{r}{v} \quad \text{and} \quad \frac{b}{d} = \frac{u}{r} = \frac{s}{v},$$

and thus $as = uc$, $br = ud$, and $uv = rs$. It follows that

$$sa^2 + rb^2 = uac + ubd = u(ac + bd) \quad \text{and}$$

$$xu^2 + xrs = xu^2 + xuv = xu(u + v) = xuy.$$

By Stewart's theorem, the quantities on the left sides of these equations are equal, and hence the right sides must also be equal. We conclude by canceling u that $ac + bd = xy$, as required. ∎

Exercises 2D

2D.1 Prove that the sum of the squares of the lengths of the medians of a triangle is three fourths the sum of the squares of the lengths of the sides.

2D.2 A quadrilateral with side lengths 1, 1, 1, and d is inscribed in a circle. Find a formula for the radius R of the circle in terms of d. Check your formula by computing R directly when $d = 1$.

   HINT: Prove that the diagonals of the quadrilateral must be equal.

2D.3 Given a regular pentagon with side length 1, compute the lengths of its diagonals.

   HINT: By Exercise 1G.2, the polygon can be inscribed in a circle. Note that all five diagonals must have the same length x and use Ptolemy's theorem to find a quadratic equation satisfied by x.

2E   The Incircle

A circle is said to be **inscribed** in a triangle if its center is interior to the triangle and all three sides of the triangle are tangent to the circle. Given an arbitrary triangle, we shall see that there must exist a unique inscribed circle. This circle is called the **incircle**; its center I is the **incenter**, and the length r of its radius is the **inradius** of the triangle. We can see informally why the incircle must exist: Start with a small circle placed inside the triangle and let it grow continuously, keeping it inside the triangle by letting its center move freely as the circle grows. Eventually, the circle will reach a maximum size, after which there is no room for further growth. At that point, the circle will be touching (that is, tangent to) all three sides.

This argument is rather hazy and unsatisfactory, and so we present a more formal proof. We begin with a lemma.

(2.24) LEMMA.   *The bisector of $\angle ABC$ is the locus of points P in the interior of the angle that are equidistant from the sides of the angle.*

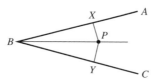

Figure 2.16

Proof.   Recall that the distance from a point to a line is measured perpendicularly. If P is in the interior of $\angle ABC$, as in Figure 2.16, drop perpendiculars PX and PY to lines AB and CB. Saying that P is equidistant from the sides of the angle, therefore, is the same as saying that $PX = PY$, and our task is to show that this happens if and only if P lies on the angle bisector.

Suppose first that $PX = PY$. Since $BP = BP$, we can conclude that $\triangle PXB \cong \triangle PYB$ by HA because these are right triangles with right angles at X and Y. It follows that $\angle XBP = \angle YBP$, and thus line BP is the bisector of $\angle ABC$. In other words, P lies on the angle bisector, as required.

Conversely now, suppose that P lies on the angle bisector. Since BP is the only line through B that contains P, it follows that BP is the angle bisector, and so $\angle XBP = \angle YBP$. Since $\angle BXP = 90° = \angle BYP$ and $BP = BP$, we see that $\triangle PXB \cong \triangle PYB$ by SAA, and we conclude that $PX = PY$, as desired.   ■

(2.25) THEOREM.   *The three angle bisectors of a triangle are concurrent at a point I, equidistant from the sides of the triangle. If we denote by r the distance from I to each of the sides, then the circle of radius r centered at I is the unique circle inscribed in the given triangle.*

Proof. Let $\triangle ABC$ be the given triangle and let I be the point where the bisectors of $\angle B$ and $\angle C$ meet. From I, drop perpendiculars IU, IV, and IW from I to sides BC, AC, and AB, respectively, and note that by Lemma 2.24, we have $IU = IW$ since I lies on the bisector of $\angle B$. Similarly, since I also lies on the bisector of $\angle C$, we see that $IU = IV$. We conclude that $IW = IV$, and thus by Lemma 2.24, point I must lie on the bisector of $\angle A$. Hence all three angle bisectors go through this point, as claimed.

We can now write $r = IU = IV = IW$, and we see that I is equidistant from sides BC, AC, and AB. Point U lies on the circle of radius r centered at I, and since BC is a line through U perpendicular to radius IU, it follows that BC is tangent to this circle at U. Similarly, AC is tangent to this circle at point V, and AB is tangent at W.

We have now shown that the circle of radius r centered at I is inscribed in $\triangle ABC$, and what remains is to prove that this is the only circle inscribed in this triangle. Suppose that we are given some inscribed circle. To prove that it is the circle we have already found, it suffices to show that its center is I and that its radius is r. Let P be the center of our unknown inscribed circle and let X, Y, and Z be the points of tangency of this circle with sides BC, AC, and AB, respectively. Then radii PX and PY are perpendicular to sides BC and AC, and since $PX = PY$, we see by Lemma 2.24 that P must lie on the bisector of $\angle C$. Similarly, P lies on the bisector of $\angle B$, and we conclude that P is the point I, as expected. It follows that each of PX and IU is a perpendicular drawn from this point to BC, and hence these are the same line segment. The radius PX of the unknown circle is thus equal to $IU = r$, and the proof is complete. ∎

In the spirit of the previous section, we ask how the inradius r can be computed and how the points of tangency U, V, and W can be located.

In the following, we refer to Figure 2.17, where we have drawn the incircle of $\triangle ABC$ and the three radii IU, IV, and IW of length r joining the center I to the three points of tangency of the circle with the sides. These radii are thus perpendicular to the corresponding sides. We have also drawn segments IA, IB, and IC, which by Theorem 2.25 we know bisect the angles of the triangle.

(2.26) PROBLEM. Given a triangle with area K, semiperimeter s, and inradius r, prove that $rs = K$. Use this to express r in terms of the lengths of the sides of the triangle.

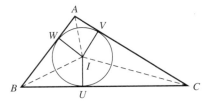

Figure 2.17

Solution.   Consider $\triangle BIC$ in Figure 2.17. Since IU is perpendicular to BC, we see that the area $K_{IBC} = \frac{1}{2}(IU)(BC) = \frac{1}{2}ra$. Similarly, $K_{IAB} = \frac{1}{2}rc$ and $K_{ICA} = \frac{1}{2}rb$. Adding these, we get

$$K = K_{IBC} + K_{IAB} + K_{ICA} = \frac{1}{2}r(a+b+c) = rs,$$

as required.

Since we know by Heron's formula (Theorem 2.18) that $K = \sqrt{s(s-a)(s-b)(s-c)}$, we deduce that

$$r = \frac{K}{s} = \sqrt{\frac{(s-a)(s-b)(s-c)}{s}},$$

which is the desired formula.

■

Next, we attempt to pin down the precise locations of the points of tangency U, V, and W along sides BC, AC, and AB. (We continue to refer to Figure 2.17.) Exercise 1F.3 asserted that the lengths of the two tangents to a circle from an exterior point are equal, and so $AV = AW$. In fact, it is easy to see this directly because $\triangle AVI \cong \triangle AWI$ by HA. We write $x = AV = AW$, and similarly, $y = BU = BW$ and $z = CU = CV$. We have $y + z = BC = a$, and similarly, $x + z = b$ and $x + y = c$. It is easy to solve these three equations for the three unknowns x, y, and z by writing $z = a - y$ from the first equation and substituting this into the second equation to get $x + a - y = b$. Thus $x - y = b - a$, and since $x + y = c$ from the third equation, we deduce that

$$x = \frac{(c+b-a)}{2} = \frac{(a+b+c)-2a}{2} = s - a$$

and similarly, $y = s - b$ and $z = s - c$.

It is interesting to note that in the calculation of the distances x, y, and z that we just completed, we did not fully use the assumption that points U, V, and W are the points of tangency of the incircle. All that was actually used were the three equations $AV = AW$, $BU = BW$, and $CU = CV$; that was enough information to determine the distances x, y, and z and thereby to determine the precise locations of the three points U, V, and W. For future reference, it is worth stating this observation as a lemma, which we have just proved.

(2.27) LEMMA.   *There is exactly one way to choose points U, V, and W on sides BC, AC, and AB, respectively, of $\triangle ABC$ so that $AV = AW$, $BU = BW$, and $CU = CV$. The only points that satisfy these equations are the points where the sides of the triangle are tangent to the incircle. Furthermore, the distances AV, BW, and CU are equal to $s - a$, $s - b$, and $s - c$, respectively, in the usual notation.*   ■

(2.28) PROBLEM.   Given three pairwise mutually externally tangent circles, show that the three common tangent lines are concurrent.

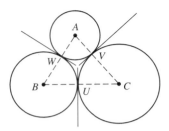

Figure 2.18

Since Problem 2.28 appears to have nothing to do with inscribed circles or even with triangles, the reader may wonder why we have presented it here. As we shall see, the key to the solution is the fact that the point of concurrence of the three tangent lines turns out to be the incenter of an appropriate triangle.

Solution to Problem 2.28.   As shown in Figure 2.18, we let U, V, and W be the three points where two of the circles touch, and we write A, B, and C to denote the centers of the three given circles. Observe that radius BU is perpendicular to the common tangent through U and that the same is true of radius CU. It follows that $\angle BUC = 180°$, and thus U lies on the line segment BC. Similarly, V lies on AC and W lies on AB.

We see now that U, V, and W are points on the sides of $\triangle ABC$, and we have $AV = AW$ since these two segments are radii of the same circle. Similarly, $BU = BW$ and $CU = CV$, and we conclude from Lemma 2.27 that points U, V, and W are the points of tangency of the incircle of $\triangle ABC$ with the sides of the triangle. It follows that radius IU of the incircle is perpendicular to side BC at U. We know, however, that the common tangent line for the circles centered at B and C is perpendicular to BC at U, and thus this common tangent must be line IU. Similarly, the other two common tangent lines also go through the incenter I of $\triangle ABC$, and hence the three common tangents are concurrent at I.   ∎

We return now to the situation of Figure 2.17, from which we can extract still more information. Given the lengths a, b, and c of the sides of $\triangle ABC$, we know that we can compute the angles of this triangle using the law of cosines. If we want $\angle C$, for example, we start with the equation $c^2 = a^2 + b^2 - 2ab\cos(C)$. From this, we deduce that $\cos(C) = (a^2 + b^2 - c^2)/2ab$, and this determines $\angle C$ since there is only one angle in the range $0°$ to $180°$ that has any given cosine. The following law of tangents is an alternative method method for finding $\angle C$.

(2.29) THEOREM (Law of Tangents).   *In $\triangle ABC$, let the quantities a, b, c, and s have their usual meanings. Then*

$$\tan\left(\frac{C}{2}\right) = \sqrt{\frac{(s-a)(s-b)}{s(s-c)}}.$$

Proof.   Consider right $\triangle CUI$ in Figure 2.17. We know that UI/UC is the tangent of $\angle UCI$, and since CI bisects $\angle C$ of the original triangle, this gives $\tan(C/2) = UI/UC = r/(s - c)$. We saw previously, however, that $r = \sqrt{(s - a)(s - b)(s - c)/s}$, and the result follows. ∎

Of course, with the obvious modifications, the law of tangents works to give formulas for $\tan(A/2)$ and $\tan(B/2)$ also. As a comparison of the laws of cosines and tangents, we use both methods to approximate $\angle C$ in a triangle for which $a = 5, b = 6$, and $c = 7$. First, using the law of cosines, we have $\cos(C) = (25 + 36 - 49)/60 = 12/60 = 1/5$. Using the inverse cosine button on a calculator, we find that $\angle C = 78.463\ldots°$. To use the law of tangents, we first compute that $s = 9$. Thus $\tan(C/2) = \sqrt{(4)(3)/(9)(2)} = \sqrt{2/3}$. Using a calculator to compute the inverse tangent of $\sqrt{2/3}$ and to double the result, we obtain the same answer for $\angle C$ as before. This calculation gives us the exact formula: $\arccos(1/5) = 2\arctan(\sqrt{2/3})$, which might be a bit tedious to try to prove directly.

It is amusing that we can use the law of tangents to produce a proof of the Pythagorean theorem. If $\angle C = 90°$, then $\tan(C/2) = 1$, and the law of tangents yields $(s - a)(s - b) = s(s - c)$. From this and the fact that $2s = a + b + c$, some algebraic manipulation, which we leave to the reader, yields that $c^2 = a^2 + b^2$, as required. In fact, this "proof" of the Pythagorean theorem is invalid because it is circular: The Pythagorean theorem underlies our proof of the law of tangents. We leave it to the reader to trace through the arguments and verify this assertion.

It is easy to see that in the case of an equilateral triangle, all of the special points we have been discussing—the circumcenter, the centroid, the orthocenter, and the incenter— coincide. It is not too hard to see that, conversely, if any two of these points coincide, then the triangle must be equilateral.

(2.30) PROBLEM.   Suppose that the incenter and the circumcenter of $\triangle ABC$ are the same point. Show that the triangle must be equilateral.

Solution.   Join O to vertices A and B and note that $\triangle OAB$ is isosceles since we know that $OA = OB$. By the pons asinorum, we have $\angle OAB = \angle OBA$. Since we are assuming that O is also the incenter, we know that OA and OB bisect angles A and B, and we conclude that $\angle A = 2\angle OAB$ and $\angle B = 2\angle OBA$. It follows that $\angle A = \angle B$, and so by the converse of the pons asinorum, we see that $CA = CB$. Since we could have started with any two of the three vertices, it follows that all of the sides are equal, as required. ∎

We now know that except when the triangle is equilateral, points I and O never coincide. It seems natural, therefore, to ask just how far apart these two points are in general. An answer is given by the following theorem of Euler.

(2.31) THEOREM (Euler).   *Let $d = OI$, the distance from the circumcenter to the incenter of an arbitrary triangle. Then $d^2 = R(R - 2r)$, where, as usual, R and r are the circumradius and inradius of the given triangle.*

We need an interesting preliminary result first.

(2.32) LEMMA. *Extend the bisector of one of the angles of a triangle to meet the circumcircle at point P. Then the distance from P to each of the other two vertices of the triangle is equal to the distance IP, where I is the incenter of the given triangle.*

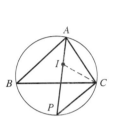

Figure 2.19

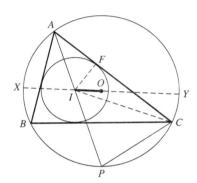

Figure 2.20

Proof. In Figure 2.19, we have drawn line AP bisecting $\angle A$ of $\triangle ABC$. Of course, I lies on this line, and our task is to show that $IP = CP$. By the converse of the pons asinorum, it suffices to show that $\angle ICP = \angle CIP$, and so we start computing angles.

Since $\angle BCP$ subtends the same arc as $\angle BAP$, we see that $\angle BCP = \angle BAP = \frac{1}{2}\angle A$. Also, IC bisects $\angle C$ of the original triangle, and thus $\angle ICB = \frac{1}{2}\angle C$, and we conclude that $\angle ICP = \frac{1}{2}(\angle A + \angle C)$.

To compute $\angle CIP$, we observe first that $\angle P = \angle B$ since these subtend the same arc. It follows that

$$\angle ICP + \angle CIP = 180° - \angle P = 180° - \angle B = \angle A + \angle C$$

and since we have seen that $\angle ICP$ is exactly half of this quantity, we deduce that $\angle CIP$ is the other half. Thus $\angle ICP = \angle CIP$, and the proof is complete. ∎

We are now ready to prove Euler's theorem.

Proof of Theorem 2.31. As shown in Figure 2.20, we extend line segment OI to a diameter XY of the circumcircle. Since $OI = d$, we see that $XI = R - d$ and $YI = R + d$, and thus $XI \cdot YI = R^2 - d^2$. By Theorem 1.35, however, we know that the product of the lengths of the two pieces of each chord through I is a constant, independent of the particular chord, and it follows that $R^2 - d^2 = XI \cdot YI = AI \cdot IP$.

Next, we try to compute the length IP, which, by Lemma 2.32, we know is equal to PC. To accomplish this, we use the extended law of sines in $\triangle APC$. Since $\angle PAC = \frac{1}{2}\angle A$, we have

$$\frac{PC}{\sin(A/2)} = 2R \,,$$

and thus

$$R^2 - d^2 = AI \cdot IP = AI \cdot PC = AI \cdot 2R\sin(A/2) \,.$$

Finally, we compute $AI \cdot \sin(A/2)$ by working in right $\triangle AIF$, where F is the point of tangency of the incircle with side AC. Since $IF = r$, we see that $\sin(A/2) = r/AI$, and thus $AI \cdot \sin(A/2) = r$. If we substitute this into our previous formula, we get $R^2 - d^2 = 2rR$, and Euler's formula follows. ∎

(2.33) COROLLARY. *For any triangle, $R \geq 2r$ and equality holds if and only if the triangle is equilateral.*

Proof. Since $d^2 = R(R - 2r)$, we see that $R - 2r$ can never be negative. Furthermore, $R = 2r$ if and only if $d = 0$; in other words, $R = 2r$ if and only if points I and O are identical. We already know, however, that I and O coincide for equilateral triangles and, by Problem 2.30, only for equilateral triangles. ∎

Exercises 2E

2E.1 Suppose that the orthocenter and incenter of $\triangle ABC$ are the same point. Prove that the triangle is equilateral.

2E.2 Suppose that the centroid and incenter of $\triangle ABC$ are the same point. Prove that the triangle is equilateral.

2E.3 Show that in a right triangle, the inradius, circumradius, and semiperimeter are related by the formula $s = r + 2R$.

2E.4 Let CD be an altitude of $\triangle ABC$ and assume that $\angle C = 90°$. Let r_1 and r_2 be the inradii of $\triangle CAD$ and $\triangle CBD$, respectively, and show that $r + r_1 + r_2 = CD$, where, as usual, r is the inradius of $\triangle ABC$.

2E.5 Extend the bisectors of $\angle A$, $\angle B$, and $\angle C$ of $\triangle ABC$ to meet the circumcircle at points X, Y, and Z. Show that I is the orthocenter of $\triangle XYZ$.

HINT: Show that XY is the perpendicular bisector of IC.

2F   Exscribed Circles

The incircle of a triangle is sometimes referred to as a **tritangent** circle because it is tangent to all three sides of the triangle. We can see in Figure 2.21 that if we are willing to consider circles tangent to *extensions* of the sides, a triangle also has three other tritangent circles. Each of these is tangent to one side and extensions of the other two sides.

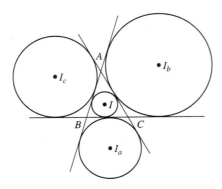

Figure 2.21

The three tritangent circles whose centers are exterior to the given triangle are called the **exscribed circles** or the **excircles** of the triangle. Although we shall not give a formal proof of the existence and uniqueness of the three exscribed circles, it should be clear by analogy with the inscribed circle how to construct such a proof. It should also be clear that, as was the case for the incenter, the center of each of the exscribed circles lies at the intersection of three angle bisectors. In Figure 2.21, for example, the center of the excircle opposite vertex A has been denoted I_a; it lies at the point of concurrence of the bisector of $\angle A$ and the bisectors of the *exterior* angles at points B and C.

The centers of the excircles of $\triangle ABC$ are the **excenters** of the triangle, and it is customary to label these points I_a, I_b, and I_c, as indicated in the figure. The corresponding radii are denoted r_a, r_b, and r_c, and they are referred to as the **exradii** of the triangle. The three exradii, together with the inradius r, are collectively known as the tritangent radii, and as it turns out, there is a pretty relationship among these four quantities.

(2.34) THEOREM.   *Given an arbitrary $\triangle ABC$, we have*

$$\frac{1}{r} = \frac{1}{r_a} + \frac{1}{r_b} + \frac{1}{r_c}.$$

Just as it was useful to locate the points of tangency of the incircle along the sides of the triangle, so too, to prove Theorem 2.34, will we need to find the points of tangency of the three excircles. We are especially interested in the external points of tangency.

(2.35) LEMMA.   *The length of the tangent from a vertex of a triangle to the opposite exscribed circle is equal to the semiperimeter s.*

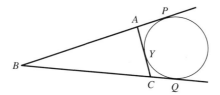

Figure 2.22

Proof.   As in Figure 2.22, let Y be the point of tangency of side AC with the excircle opposite B and let P and Q be the points of tangency of this circle with the extensions of AB and BC. Since the two tangents to a circle from an exterior point are equal, we know that $BP = BQ$, and we need to show that this common length is s.

Since $AP = AY$, we have $BP = BA + AP = BA + AY$, and similarly, $BQ = BC + CY$. Adding these equations, we obtain

$$BP + BQ = BA + AY + BC + CY = BA + BC + AC = 2s \, .$$

But BP and BQ are equal, and since their sum is $2s$, it follows that each of them is equal to s.   ∎

We can use Lemma 2.35 to express the exradii of $\triangle ABC$ in terms of the side lengths a, b, and c. Observe that excenter I_b lies on the bisector of $\angle B$. If, as in Figure 2.22, we write P to denote the point of tangency of the excircle opposite B with the extension of side AB, we see from Lemma 2.35 that the $\triangle BPI_b$ is a right triangle with arm BP of length s. Since arm I_bP has length r_b, it follows that $\tan(B/2) = r_b/s$. By the law of tangents, we obtain

$$r_b = s \tan\left(\frac{B}{2}\right) = s\sqrt{\frac{(s-a)(s-c)}{s(s-b)}} = \sqrt{\frac{s(s-a)(s-c)}{s-b}} \, .$$

Proof of Theorem 2.34.   We have

$$\frac{1}{r_a} + \frac{1}{r_b} + \frac{1}{r_c} = \sqrt{\frac{s-a}{s(s-b)(s-c)}} + \sqrt{\frac{s-b}{s(s-a)(s-c)}} + \sqrt{\frac{s-c}{s(s-a)(s-b)}} \, .$$

Combining these over a common denominator, we see that the above sum is equal to

$$\frac{(s-a) + (s-b) + (s-c)}{\sqrt{s(s-a)(s-b)(s-c)}} = \frac{s}{\sqrt{s(s-a)(s-b)(s-c)}} = \frac{s}{K} = \frac{1}{r}$$

by Problem 2.26.   ∎

Exercises 2F

2F.1 In Figure 2.22, show that $AY = s - c$ and $CY = s - a$.

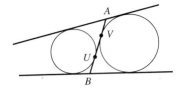

Figure 2.23

2F.2 In Figure 2.23, three common tangent lines have been drawn to two circles. Show that $AV = BU$.

HINT: View the two circles as the incircle and one of the excircles of an appropriate triangle.

2G   Morley's Theorem

We devote this section to an amazing theorem discovered by Frank Morley at the beginning of the 20th century. Although this result is somewhat awkward to state in words, the diagram of Figure 2.24 makes the assertion strikingly clear.

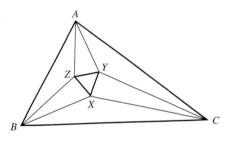

Figure 2.24

(2.36) THEOREM (Morley).   *Draw the six angle trisectors for an arbitrary triangle and for each side of the triangle, mark the intersection point of the two trisectors nearest that side. Then the triangle formed by the three marked points is equilateral.*

In the figure, lines AY and AZ divide $\angle A$ into three equal parts. Similarly, BX and BZ trisect $\angle B$, and CX and CY trisect $\angle C$. Morley's theorem says that $\triangle XYZ$ must be equilateral.

We need a preliminary result about incenters.

(2.37) LEMMA.   *Let AX be the bisector of $\angle A$ in $\triangle ABC$, where X lies on side BC. Then the incenter of $\triangle ABC$ is the unique point P on segment AX such that $\angle BPC = 90° + \frac{1}{2}\angle BAC$.*

Proof. First, assume that P is the incenter. Then P lies on all three angle bisectors, and we compute in $\triangle BPC$ that

$$\angle BPC = 180° - \angle PBC - \angle PCB = 180° - \frac{\angle B}{2} - \frac{\angle C}{2}$$

$$= 180° - \frac{1}{2}(180° - \angle A) = 90° + \frac{1}{2}\angle A,$$

and thus the incenter has the property that was claimed.

To see that the incenter is the only point on AX with the stated property, consider what happens to $\angle BPC$ as point P moves along segment AX from point A to point X. It is clear from a diagram that $\angle BPC$ is monotonically strictly increasing from a minimum equal to $\angle A$ when P is at A to a maximum of $180°$ when P is at X. It follows that there can be just one point P where $\angle BPC$ takes on any particular value, and thus the incenter is the only point where $\angle BPC = 90° + \frac{1}{2}\angle A$.   ∎

We shall use a rather unusual method to prove Morley's theorem. So before we begin, it might be a good idea to explain the strategy. We shall describe the diagram of Figure 2.24 as the **Morley configuration** associated with $\triangle ABC$. By this, we mean that the six lines AY, AZ, BX, BZ, CX, and CY are the trisectors of $\angle A$, $\angle B$, and $\angle C$. Our goal, of course, is to show that for every Morley configuration, $\triangle XYZ$ is equilateral, and our proof will proceed by constructing a large variety of Morley configurations for which $\triangle XYZ$ actually is equilateral. (We shall say that a Morley configuration is **successful** if $\triangle XYZ$ is equilateral.) But how does the construction of successful configurations prove the theorem?

The key here is to observe that if we start with a pair of similar triangles, then each differs from the other by expansion or contraction via an appropriate scale factor. It follows that the entire Morley configurations associated with these two triangles differ from one another by expansion or contraction via the same scale factor, and thus if one is successful, then so is the other. Given an arbitrary triangle, therefore, it suffices to produce a successful Morley triangle associated with some triangle similar to the given triangle.

Proof of Theorem 2.36. To guarantee that the Morley configuration we produce is successful, we work backward. We *start* with equilateral $\triangle XYZ$, as shown in Figure 2.25, and we work to construct a larger $\triangle ABC$, which together with $\triangle XYZ$

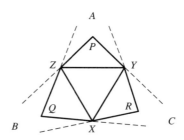

Figure 2.25

will form a Morley configuration. We will be done if we can do this in such a way that $\triangle ABC$ will be similar to the given unshown $\triangle UVW$. Let α, β, and γ be $\frac{1}{3}\angle U$, $\frac{1}{3}\angle V$, and $\frac{1}{3}\angle W$, respectively, and note that $\alpha + \beta + \gamma = \frac{1}{3}180° = 60°$.

Build isosceles $\triangle ZPY$ on side YZ of $\triangle XYZ$, as shown in Figure 2.25, and do this so that $\angle PZY = \angle PYZ = \beta + \gamma$. To see that this is possible, it suffices to observe that $\beta + \gamma < 90°$. Similarly, construct points Q and R so that $\triangle XQZ$ is isosceles with $\angle QZX = \angle QXZ = \alpha + \gamma$ and $\triangle XRY$ is isosceles with $\angle RXY = \angle RYX = \alpha + \beta$.

Next, extend segment QZ in the direction of Z and extend RY in the direction of Y. We claim that these extensions meet at some point A in the vicinity of the label A in Figure 2.25. To see why this is so, it suffices to consider the sum $\theta = \angle QZY + \angle RYZ$. Observe that if $\theta = 180°$, that would imply that $QZ \| RY$. Otherwise, lines QZ and RY meet, and we observe that if $\theta < 180°$, then the intersection point would lie below line YZ in the figure, and if $\theta > 180°$, then the intersection point is above YZ, as we have claimed. In fact, $\angle QZY = \angle QZX + \angle XZY = \alpha + \gamma + 60°$, and similarly, $\angle RYZ = \alpha + \beta + 60°$. Since $\alpha + \beta + \gamma = 60°$, it follows that

$$\theta = 2\alpha + \beta + \gamma + 120° = \alpha + 180°.$$

This exceeds $180°$, and we conclude that the intersection point A of lines QZ and RY is positioned as indicated, above line YZ. Also, working in $\triangle AZY$, we see that $\angle A = 180° - \angle AZY - \angle AYZ$, and thus

$$\angle A = 180° - (180° - \angle QZY) - (180° - \angle RYZ) = \theta - 180° = \alpha.$$

Similarly, lines PZ and RX meet at B, and lines QX and PY meet at C, where points B and C are positioned as indicated in Figure 2.25. Also, we have $\angle B = \beta$ and $\angle C = \gamma$.

We now work to show that Z is the incenter of $\triangle ABR$. Observe first that $\triangle ZYR \cong \triangle ZXR$ by SSS, and thus $\angle YRZ = \angle XRZ$, and Z lies on the bisector of $\angle ARB$. If we can show that $\angle AZB = 90° + \frac{1}{2}\angle ARB$, it will follow by Lemma 2.37 that Z is the incenter of $\triangle ARB$, as desired. We have

$$\angle AZB = \angle PZQ = \angle PZY + \angle YZX + \angle XZQ = (\beta + \gamma) + 60° + (\alpha + \gamma)$$
$$= 120° + \gamma,$$

since $\alpha + \beta + \gamma = 60°$. Also, we see in $\triangle YRX$ that

$$\angle R = 180° - \angle RYX - \angle RXY = 180° - 2(\alpha + \beta) = 180° - 2(60° - \gamma)$$
$$= 60° + 2\gamma,$$

and thus

$$90° + \frac{1}{2}\angle R = 90° + 30° + \gamma = 120° + \gamma = \angle AZB,$$

as required. It follows that Z is the incenter of $\triangle ARB$, as claimed, and thus AZ bisects $\angle BAR$, and we have $\angle BAZ = \angle ZAY = \alpha$. Similarly, Y is the incenter of

$\triangle AQC$, and we deduce that $\angle YAC = \alpha$. Thus $\angle BAC = 3\alpha$, and AY and AZ are the trisectors of $\angle A$ in $\triangle ABC$.

Similar reasoning shows that $\angle B = 3\beta$ and BZ and BX are the trisectors of this angle in $\triangle ABC$. Also $\angle C = 3\gamma$ and CY and CX are angle trisectors, and thus we have the Morley configuration associated with $\triangle ABC$. In fact, this is a successful configuration since we started with $\triangle XYZ$ being equilateral. Furthermore, $\triangle ABC$ is similar to our given $\triangle UVW$ by AA, and the proof is complete. ∎

Exercises 2G

2G.1 Join each vertex of $\triangle ABC$ to the points of trisection of the opposite side and let X, Y, and Z be the points of intersection of these side-trisectors, as shown in Figure 2.26. Show that the sides of $\triangle XYZ$ are parallel to corresponding sides of $\triangle ABC$ and that $\triangle XYZ \sim \triangle ABC$.

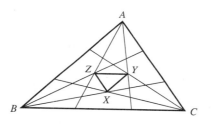

Figure 2.26

2H   Optimization in Triangles

Our goal in this section is to solve certain geometric optimization, or max-min, problems by geometric methods. As an example of what we mean, we begin with something easy.

(2.38) PROBLEM. Given points P and Q on the same side of a given line, find the point X on the line that minimizes the total distance $PX + XQ$. Show that at this

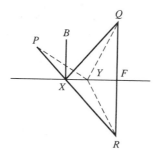

Figure 2.27

point, the bisector of $\angle PXQ$ is perpendicular to the line and that PX and QX make equal angles with the line.

Solution.   First, reflect one of the given points (say, Q) in the given line to get R, as in Figure 2.27. Recall that this means that we drop a perpendicular QF from Q to the line and then choose R on the line QF, on the opposite side of the given line from Q, so that $QF = FR$. Next, draw PR and let X be the point where PR meets the given line.

   We claim that this point X is the unique solution to our optimization problem. To establish this, we must prove that $PX + XQ < PY + YQ$ for every possible choice of a point Y on the line, with Y different from X. To prove the inequality, we observe that $XQ = XR$ and $YQ = YR$ since line XY is the perpendicular bisector of segment QR. Thus

$$PX + XQ = PX + XR < PY + YR = PY + YQ$$

as required, where the inequality holds since $PX + XR = PR$ is the straight-line distance from P to R, and so is shorter than the broken-line distance $PY + YR$ from P to R. Note that we could also have found a solution by reflecting P rather than Q. This would necessarily have yielded the same point X because we have seen that X is the unique solution to the problem.

   We must show that the bisector XB of $\angle PXQ$ is perpendicular to the given line. To see this, note that $\angle PXQ$ is an exterior angle of $\triangle QXR$, and thus $\angle PXQ = \angle R + \angle Q = 2\angle R$, where the last equality follows via the pons asinorum from the fact that $XQ = XR$. Thus $\angle PXB = \frac{1}{2}\angle PXQ = \angle R$, and it follows that $XB \| RQ$. Since QR is perpendicular to the given line, so too is XB. Finally, to see that PX and QX make equal angles with the line, observe that these two angles are just the complements of $\angle PXB$ and $\angle QXB$, which are equal.   ∎

Next we consider a harder and more interesting problem.

(2.39) PROBLEM.   Given $\triangle ABC$, find points X, Y, and Z on sides BC, AC, and AB, respectively, such that the perimeter of $\triangle XYZ$ is minimized.

First, let us consider whether or not Problem 2.39 necessarily has a solution. If we wish to prove the existence of a minimum using the compactness technique that was discussed in Section 1G, we need to be sure that the domain of choice for each of the variables X, Y, and Z is closed. This will be the case if we interpret the requirement in the problem that the points are chosen on the sides of $\triangle ABC$ to allow the possibility that one or more of the points are actually at a vertex. If we do that, however, we can no longer be sure that X, Y, and Z are distinct points, and in that case, we must agree about the meaning of the perimeter of the "triangle" formed by these points. Our interpretation is that the perimeter of $\triangle XYZ$ is the sum of the distances XY, YZ, and ZX, even if one or more of these distances is zero. With this interpretation, and allowing X, Y, or Z to be at a vertex, Problem 2.39 always has a solution.

We shall see that if one of the angles (say, $\angle A$) of the original triangle is not acute, then the minimum perimeter for $\triangle XYZ$ occurs when both Y and Z are at A. It is clear in this case that X must be at the foot of the altitude from A and that the minimum perimeter is $2AX$. If we had decided to interpret the problem strictly and we did not allow Y and Z to coincide with A, there would be no solution in this case. This is because wherever we place Y and Z, we can always reduce the perimeter of $\triangle XYZ$ by moving these points closer to A and adjusting X appropriately.

The key to the solution of Problem 2.39 is the following lemma.

(2.40) LEMMA.   *Given $\triangle ABC$ with acute angles at B and C and $\angle A = \theta$, fix a point P on side BC and let Q and R be the reflections of P in sides AB and AC, respectively. Let U and V be variable points on sides AB and AC and consider the perimeter p of $\triangle UPV$.*

  a. *If $\theta < 90°$, then the minimum value p_{min} of p occurs when and only when U and V coincide with the points Z and Y where line QR meets sides AB and AC. Furthermore, $p_{min} = QR = 2AP \cdot \sin(\theta)$.*

  b. *If $\theta \geq 90°$, then the minimum value of p occurs when and only when U and V both coincide with point A.*

Proof.   Suppose first that $\theta < 90°$ and refer to Figure 2.28. Using congruent triangles, it is easy to see that $\angle QAB = \angle BAP$ and $\angle CAR = \angle PAC$. It follows that

$$\angle QAR = \angle QAB + \angle BAP + \angle PAC + \angle CAR = 2(\angle BAP + \angle PAC) = 2\theta < 180°,$$

and thus segment QR really does intersect AB and AC, as shown in Figure 2.28. Note that if $\theta > 90°$, this calculation shows that segment QR passes above point A and that if $\theta = 90°$, then QR passes through A.

Since AB is the perpendicular bisector of segment QP, we see that $UP = UQ$, and similarly, $VP = VR$. It follows that $p = PU + UV + VP = QU + UV + VR$ is a broken-line distance from Q to R unless U coincides with Z and V coincides with Y, in which case $p = QR$ is the straight-line distance from Q to R. It follows that p is minimized when U and V are at Z and Y and that every other possible choice of U and V yields $p > QR$.

To compute the length QR, consider $\triangle AQR$. We know that $AQ = AP = AR$, and so this triangle is isosceles. If we let M be the midpoint of QR, then

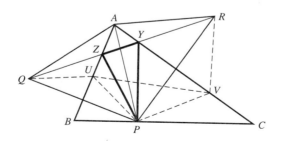

Figure 2.28

AM is perpendicular to QR and AM bisects $\angle QAR$. Thus $\angle QAM = \theta$ and $\sin(\theta) = QM/AQ$. It follows that $QR = 2QM = 2AQ \cdot \sin(\theta) = 2AP \cdot \sin(\theta)$. This completes the proof of part (a).

Assume now that $\angle BAC \geq 90°$. As we have seen, line QR passes above or through point A in this case, and it is clear from a picture that the shortest path from Q to U to V to R with U on AB and V on AC occurs when U and V are both at A.   ■

Solution to Problem 2.39.   If the given triangle is not acute angled, we can assume that $\angle A \geq 90°$. Wherever we place point X on side BC, we know from Lemma 2.40 that the optimum position for Y and Z so as to minimize the perimeter p is to have them both coincide with A. In this situation, $p = 2AX$, and so we can minimize p by minimizing AX. We accomplish this by placing X at the foot of the altitude from A.

Much more interesting is the case where all angles of $\triangle ABC$ are acute, and we state the result in that case as a theorem.   ■

(2.41) THEOREM.   *The pedal triangle of a given acute angled triangle has a smaller perimeter than any other triangle whose vertices lie on the three sides of the given triangle.*

Proof.   For any choice of point X on side BC, we know by Lemma 2.40 that we can choose Y on AC and Z on AB so that the perimeter of $\triangle XYZ$ is equal to $2AX \cdot \sin(A)$, and this is the smallest possible perimeter for the given point X and for any choice of Y and Z. To find the overall minimum perimeter, therefore, we must choose X so as to minimize AX, since $\sin(A)$ is constant. The smallest possible perimeter occurs, therefore, when X is the foot of the altitude from A, and Y and Z are as specified by Lemma 2.40. No other choice of the point X allows for a perimeter as small as this, no matter how points Y and Z are selected.

We have seen that there is a minimum possible perimeter for $\triangle XYZ$ and that it can occur only when X is at the foot of the altitude from A. Similar reasoning shows that for $\triangle XYZ$ to have the minimum perimeter, Y and Z must be the feet of the altitudes from B and from C. It follows that among all possibilities for $\triangle XYZ$, the pedal triangle, and only the pedal triangle, has the minimum perimeter.   ■

There is more information available from our analysis. Given acute angled $\triangle ABC$ with altitude AX, we know from Lemma 2.40 that to minimize the perimeter of $\triangle XYZ$ with Y on AC and Z on AB, we must take points Y and Z on line QR, where Q and R are the reflections of point X in sides AB and AC. On the other hand, Theorem 2.41 tells us that if we take Y and Z to be the other two altitude feet, then this will minimize the perimeter of $\triangle XYZ$. Combining these facts, we deduce that altitude feet Y and Z lie on QR. This proves the following.

(2.42) COROLLARY.   *In an acute angled triangle, the reflection of the foot of an altitude in either of the sides not containing it is collinear with the other two altitude feet.*   ∎

We can get still more information from Theorem 2.41. Fix altitude feet Y and Z on sides AC and AB of acute angled $\triangle ABC$ and consider the point X on line BC that minimizes the distance $ZX + XY$. We know from Problem 2.38 that there is such a point. It is clear from a diagram that X must lie between B and C. Also, because we are holding Y and Z fixed, we see that minimizing $ZX + XY$ is exactly equivalent to the problem of choosing X to minimize the perimeter of $\triangle XYZ$. Since Y and Z are two of the vertices of the pedal triangle, it follows by Theorem 2.41 that the minimum perimeter will occur when X coincides with the third vertex of the pedal triangle.

We have now shown that if X is chosen on side BC so as to minimize $YX + XZ$, then AX is an altitude of the given acute angled $\triangle ABC$. We know from Problem 2.38, however, that for this point X, the bisector of $\angle YXZ$ is perpendicular to BC. But AX is an altitude of $\triangle ABC$, and so it is perpendicular to BC at X, and we conclude that altitude AX bisects $\angle YXZ$. We have proved the following.

(2.43) COROLLARY.   *The altitudes of an acute angled triangle are the angle bisectors of its pedal triangle.*   ∎

We consider next a problem proposed by Pierre Fermat (1601–1665).

(2.44) PROBLEM (Fermat).   Locate the point F in the interior of triangle ABC that minimizes $FA + FB + FC$.

Since the interior of a triangle is not a compact set, we have no guarantee that Problem 2.44 has a solution. If we reinterpret the problem, however, and allow the possibility that F lies either inside the triangle or on one of its sides or vertices, then a solution must exist. It turns out that if one of the angles (say, $\angle A$) of the given triangle is 120° or larger, then the solution to the modified problem is to take F to be A. If we really insist that F be interior to the triangle, there is no solution in this case. As we shall see, there is a very pretty solution to Fermat's problem when all three angles of $\triangle ABC$ are less than 120°. In that case, it turns out that there is a unique point F in the interior of the triangle (called the **Fermat point**) for which the quantity $FA + FB + FC$ is minimized. We shall see that this point has some additional amazing properties: It is the point of concurrence of three interesting lines and three interesting circles.

(2.45) LEMMA.   *Build an outward-pointing equilateral $\triangle ABR$ on side AB of $\triangle ABC$.*
  a. *For every point X in the plane, we have $XA + XB + XC \geq RC$. Equality can occur only when X lies on line segment RC.*
  b. *If all angles of $\triangle ABC$ are less than 120°, let F be the point, other than R, where line RC meets the circumcircle of $\triangle ABR$. Then F lies inside $\triangle ABC$ and $FA + FB + FC = RC$.*

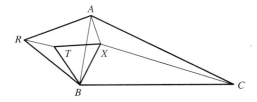

Figure 2.29

Proof.   We begin by constructing equilateral $\triangle XBT$, as shown in Figure 2.29. To be more precise, we draw line segment BX; we rotate it through an angle of 60°, with point B as the center of rotation, to obtain segment BT, and then we draw XT to complete the equilateral triangle. Furthermore, we specify that the direction of the 60° rotation of BX is the same as the direction through which one would have to rotate BA by 60° to get BR. Observe that since each possible position for X yields an unambiguous determination of T, we can think of T as a function of the variable point X. In particular, if X is at A, then T is at R, and if X is at B, then T is also at B, and in this case, $\triangle XBT$ degenerates to a single point.

We claim that $\triangle AXB \cong \triangle RTB$. The easiest way to see this is to observe that a 60° rotation about point B carries point A to point R, and the same rotation carries point X to point T. Since this rotational transformation clearly carries point B to itself, the net effect of the transformation is to move $\triangle AXB$ so as to make it coincide with $\triangle RTB$, and thus the triangles are congruent. A more traditional argument using SAS is also available since $BA = BR$, $BX = BT$, and it is easy to see that $\angle ABX = \angle RBT$.

We now have $XA = RT$. We also know that $XB = XT$, and thus $XA + XB + XC = RT + TX + XC$. The length of this path from R to C cannot be less than the straight-line distance RC. In fact, it cannot even be equal to RC unless point X (and also point T) lies on segment RC. This completes the proof of part (a).

It is not hard to see in Figure 2.30 that if $\angle A < 120°$ and $\angle B < 120°$, then line RC crosses line AB between points A and B. Note that if $\angle A = 120°$, then point R would lie on an extension of side AC and RC would cut AB at A; similarly, if $\angle B = 120°$, then RC would cut AB at B.

Continuing to assume that $\angle A$ and $\angle B$ are less than 120°, we draw the circumcircle of equilateral $\triangle ARB$ and define F to be the point other than R

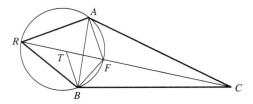

Figure 2.30

where line RC cuts the circle. Assuming also that $\angle C < 120°$, we claim that point F lies inside $\triangle ABC$, as illustrated in Figure 2.30. To see this, observe that $\angle AFB \overset{\circ}{=} \frac{1}{2} \overset{\frown}{ARB} = \frac{1}{2}(240°) = 120°$. Since $\angle C < \angle AFB$, it follows that point C lies outside the circle, as shown, and thus F lies inside the triangle.

In this case, where all three angles of $\triangle ABC$ are less than $120°$, we take the point X of the first part of the proof to be F, and we argue that the corresponding point T lies on segment RF, as shown in Figure 2.30. To see that this is true, we first observe that $\angle RFB \overset{\circ}{=} \frac{1}{2} \overset{\frown}{RB} = 60°$, and hence T lies somewhere on line RC, to the left of point F in the diagram. To see that T lies between F and R, it suffices to show that $\angle TBF < \angle RBF$. But this is clear since $\angle TBF = 60°$ and $\angle RBF > \angle RBA = 60°$.

We know from the first part of the proof that $FA + FB + FC = RT + TF + FC$. When all three angles of the original triangle are less than $120°$, we have seen that the "broken line" $RTFC$ is actually straight, and the latter sum is thus exactly equal to RC. This completes the proof of part (b). ∎

As promised, we can now solve Fermat's problem (Problem 2.44) in the case where none of the angles of $\triangle ABC$ is as large as $120°$. Some of the amazing properties of the Fermat point are visible in Figure 2.31, but Theorem 2.46 tells us that even more is true.

(2.46) THEOREM. *Given a triangle with all three angles less than 120°, construct outward-facing equilateral triangles on each of the sides. Then the line segments joining the vertices of the given triangle with the far vertices of the equilateral triangles on the opposite sides have equal lengths, and they are concurrent at some point F inside the original triangle. Also, the quantity FA + FB + FC is equal to the common length of the three segments and is smaller than XA + XB + XC for any other point X in the plane. In addition, the point F lies on the circumcircles of all three equilateral triangles.*

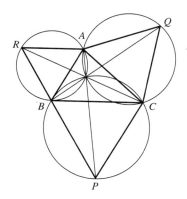

Figure 2.31

Proof.   We label points as in Figure 2.31 so that it is line segments PA, QB, and RC that we must prove are concurrent and of equal length. Define F as in Lemma 2.45(b) to be the point where RC meets the circumcircle of $\triangle ABR$ in the interior of $\triangle ABC$. By the lemma, we know that $FA + FB + FC = RC$ and that this quantity is the minimum possible value that $XA + XB + XC$ can have as X runs over all points in the plane. It follows similarly that QB is also the minimum possible value of $XA + XB + XC$, and thus, since there is at most one minimum possible value, we must have $QB = RC$. Similarly, $AP = RC$, and so the three line segments have equal lengths, as required.

We now know that $FA + FB + FC = QB$, and so by Lemma 2.45(a) applied to Q in place of R, it follows that F must lie on line segment QB. Similarly, F lies on AP, and the three segments are concurrent at F. This argument also shows that if X is any point for which the quantity $XA + XB + XC$ is minimized, then X must also lie on all three segments, and so X is the point F. In other words, F is the unique point where the minimum is attained.

This reasoning shows that the point of concurrence of PA, QB, and RC lies on the circumcircle of $\triangle ABR$. It follows similarly that this point lies on the other two circumcircles, and the proof is complete.   ∎

(2.47) COROLLARY.   *In the situation of Theorem 2.46, the six angles formed by the three concurrent line segments are all equal to* 60°.

Proof.   Observe in Figure 2.31 that $\angle AFR \overset{\circ}{=} \frac{1}{2} \overset{\frown}{AR} = 60°$, where F is the point of concurrence. The corollary follows by applying similar reasoning to the remaining five angles.   ∎

If F is the Fermat point of $\triangle ABC$, which has all angles smaller than 120°, it follows from Corollary 2.47 that $\angle AFB = \angle BFC = \angle CFA = 120°$. There is also a "physics proof" of this fact. Imagine that the plane of $\triangle ABC$ is the horizontal surface of a table and that tiny holes have been bored through the tabletop at points A, B, and C. Strings are threaded through the three holes and tied together above the table. Equal weights are attached to the lower ends of the three strings, and the system is then allowed to come into equilibrium. (The table is high enough so that none of the weights touches the floor.) The knot, which we assume is just a single point, comes to rest at some point on the plane, and we attempt to locate this point. It is perhaps not obvious that there is just one such equilibrium point, but we shall see that, in fact, this is the case.

The potential energy of the system is proportional to the sum of the heights above the ground of the three weights, and the system is in stable equilibrium precisely when this energy is at a local minimum, or equivalently, when the total length of string below the tabletop is at a local maximum. In other words, the knot is in stable equilibrium at some point X precisely when the quantity $XA + XB + XC$, which is the total length of string above the table, is at a local minimum. In particular, the Fermat point F is one possible resting place for the knot. When the knot comes to rest, there is no net force acting on it, and so the three force vectors corresponding to the tension in the three strings must sum to zero. The magnitudes of these vectors are equal, however, since the three weights are equal, and from this it easily follows that the three angles between the strings are each 120°, as we wanted to show.

In Figure 2.31, we see that the locus of points X inside $\triangle ABC$ such that $\angle AXB = 120°$ is exactly arc $\overset{\frown}{AB}$ of the circumcircle of equilateral $\triangle ABR$. It follows in our physics experiment that every possible equilibrium position for the knot lies on this arc, and similarly, all equilibrium positions must also lie on the other two circumcircles. It follows from this that the Fermat point, which is where the three circles meet, is the unique equilibrium point, and thus the function $XA + XB + XC$ has no local minimum other than the global minimum that occurs when X is at the Fermat point F.

In Chapter 5, we establish yet another remarkable fact about the configuration of Figure 2.31: The centers of the three circles form an equilateral triangle, a theorem sometimes attributed to Napoleon Bonaparte.

We close by mentioning a remarkable generalization of part of Theorem 2.46. Suppose that $\triangle ABC$ is completely arbitrary, with no restriction on its angles, and build outward-pointing isosceles $\triangle ABR$, $\triangle BCP$, and $\triangle CAQ$ having bases AB, BC, and CA. We replace the assumption of Theorem 2.46 that these triangles are equilateral with the weaker condition that all of the base angles are equal, but not necessarily equal to 60°, as in Theorem 2.46. The amazing fact is that even in this generality, it is still true that lines AP, BQ, and CR are concurrent. A proof of this is presented in Chapter 4.

Exercises 2H

2H.1 A rectangle has side lengths a and b, and as shown in Figure 2.32, points P and Q are selected on the sides of length a so that PQ is parallel to the sides of length b. Show that there exist uniquely determined points X and Y on the sides of length b such that $PX + XY + YQ$ is minimized and show that the minimum possible value for this quantity is $\sqrt{4a^2 + b^2}$.

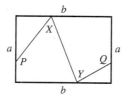

Figure 2.32

CHAPTER THREE

Circles and Lines

3A Simson Lines

The next several topics really have little in common except that, as with much of geometry, they involve circles and lines. Our first result, which is a theorem attributed to Robert Simson (1687–1768), establishes the existence of certain lines associated with the circumcircle of a triangle. As we shall see, there is also a connection between these Simson lines and the nine-point circle of the triangle, and so this section provides a continuation of some of the material from the previous chapter. Our concern here is the configuration illustrated in Figure 3.1.

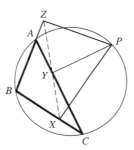

Figure 3.1

Given $\triangle ABC$, choose an arbitrary point P on its circumcircle and drop perpendiculars PX, PY, and PZ to sides BC, AC, and AB, as shown. It is almost always necessary to extend at least one of the sides of the triangle to do this. In Figure 3.1, for example, we had to extend side AB to meet the perpendicular from P. Simson's theorem asserts the amazing fact that the feet X, Y, and Z of the three perpendiculars from point P are always collinear. Appropriately, the line through X, Y, and Z is called the **Simson line** of $\triangle ABC$ with respect to point P, and P is referred to as the **pole** for this Simson line.

Before we prove Simson's theorem, let us consider the "degenerate" case where the pole P is chosen to be one of the vertices of the triangle. Suppose P is at A, for example. Then X is clearly the foot of the altitude from A in $\triangle ABC$, but where are Y

and Z in this case? The line perpendicular to AC through P, which is A, meets AC at A, and thus Y is the point A, and similarly, Z is also at A. In this situation, points X, Y, and Z are really just two points, and so they are certainly collinear. The Simson line of a triangle with respect to a pole at one of its vertices, therefore, is just the altitude of the triangle from that vertex, and in particular, the Simson line goes through its own pole in that case. It is not hard to show that, in fact, the only way that a Simson line can go through its pole is if the pole is at one of the vertices of the triangle. This is also the only way that two of the points X, Y, and Z can coincide, and also, it is only in this case that it might not be necessary to extend one of the sides of the triangle to construct these points.

The following result is the key to our proof of Simson's theorem, but it also yields some useful additional information.

(3.1) THEOREM. *Choose a point P on the circumcircle of $\triangle ABC$ and let Q be the other point where the perpendicular to BC through P meets the circumcircle. Let X be the point where this perpendicular meets line BC and let Z be the point where the perpendicular to AB through P meets AB. If Q is different from A, then Z lies on the line parallel to QA through X.*

Figure 3.2 shows two of the several possible configurations that can occur in Theorem 3.1. In the diagram on the left, we chose point P so that the foot X of the perpendicular from P to side BC falls on an extension of that side. On the right, we started with the "same" triangle, but here P was chosen so that X actually lies on *segment* BC. In each of these diagrams, point Z falls on side AB, but it can also happen that AB has to be extended to construct Z. To see an example of this, consider the right diagram, but exchange the labels B and C so that AB is now the nearly vertical side of the triangle, and the new Z is not on this line segment. Note that X and Q are not affected by the relabeling of B and C, and so the theorem tells us that the new point Z also lies on the line through X parallel to AQ. Since the original Z also lies on this line, the new Z, the original Z, and X are collinear. As we shall see, this is how Simson's theorem is proved.

Before we proceed with the proof of Theorem 3.1, we should mention a possible degeneracy. Although Q was defined as the "other" point where the perpendicular to BC

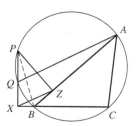
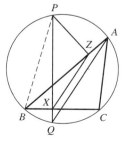

Figure 3.2

through P meets the circle, it can happen that this perpendicular is tangent to the circle, and in that case, there is no second point of intersection. The theorem is still true in this situation if we take Q to be P, and the proof will go through with only slight change. We leave the verification of this to the exercises.

Proof of Theorem 3.1.   If X and Z happen to be the same point, there is really nothing to prove. So we can assume that X and Z are distinct, and our goal is to prove that $XZ \parallel QA$. If P is at the point B, then X and Z are also at B, and since we are assuming that X and Z are different, this does not happen. Thus P and B are different, and we can consider the unique circle having diameter PB. Since $\angle PXB = 90° = \angle PZB$, it follows that points X and Z lie on this circle, and thus $\angle PXZ = \angle PBZ$ since these angles subtend the same arc in that circle. But assuming that P is different from Q, we see that $\angle PBZ = \angle PQA$ because these angles subtend the same arc in the original circle. Thus $\angle PXZ = \angle PQA$, and it follows that XZ and QA are parallel, as desired.   ■

In the statement of Theorem 3.1, it was necessary to assume that Q is different from A, since otherwise, the line QA would not be defined, and the theorem would make no sense. To see how the assumption that Q and A are different could possibly fail, suppose that these points actually do coincide. Then AX is the line QX, which is perpendicular to BC, and thus QX is the altitude from A in $\triangle ABC$. But P lies on QX, and thus P lies on the altitude from A. In other words, if P is not on the altitude from A, then the assumption in Theorem 3.1 that Q and A are distinct is guaranteed to hold.

To prove Simson's theorem, we need to consider the relationship between the orthocenter and the circumcircle of a triangle. If all of the angles of $\triangle ABC$ are acute, then we know that the orthocenter lies inside the triangle, and hence it lies inside the circumcircle. But if one of the angles of the triangle is obtuse, then it is easy to see from a diagram that the orthocenter lies outside of the circumcircle. Finally, for a right triangle, the orthocenter coincides with a vertex, and so it lies on the circumcircle.

(3.2) THEOREM (Simson's Theorem).   *Let P be any point on the circumcircle of $\triangle ABC$ and let X, Y, and Z be the feet of the perpendiculars dropped from P to lines BC, AC, and AB, respectively. Then points X, Y, and Z are collinear.*

Proof.   Suppose first that P does not lie on the altitude from A in $\triangle ABC$ and let Q be as in Theorem 3.1. Since P is not on the altitude from A, we have seen that the hypothesis in Theorem 3.1 that Q and A are distinct points is guaranteed to hold. By Theorem 3.1, therefore, we know that Z lies on the line through X parallel to AQ. Exactly similar reasoning shows that Y also lies on the line through X parallel to AQ, and thus X, Y, and Z lie on a common line, as required.

We have now shown that X, Y, and Z are collinear if P does not lie on the altitude from A. By similar reasoning, we get the same conclusion if P fails to lie on the altitude from B or if it fails to lie on the altitude from C. The only case in which we have not yet proved the theorem, therefore, is when P lies on all three altitudes, in which case P is the orthocenter of $\triangle ABC$. But P lies on the circumcircle of this triangle, and we have seen that it is only for a right triangle that the orthocenter can lie on the circumcircle.

We can now assume that $\triangle ABC$ is a right triangle and that P is its orthocenter. We can suppose that $\angle B$ is the right angle, and it follows that P is at B, and thus X and Z are also at B. Since X and Z are the same point in this case, the points X, Y, and Z are certainly collinear, and the proof is complete. ∎

(3.3) PROBLEM. Where on the circumcircle of $\triangle ABC$ should we take the pole P so that the corresponding Simson line will be parallel to one of the sides of the triangle?

Solution. It may not be obvious that it is possible to find such a pole, but let us analyze the situation. For definiteness, suppose that we want a Simson line parallel to side BC. By definition, every Simson line contains at least one point on line BC, and so if a Simson line is parallel to BC, this Simson line must actually be BC. We ask, therefore, if it is possible to find a pole P for which the corresponding Simson line is the line BC. If such a point exists and we draw the perpendicular from P to side AC, then the foot Y of this perpendicular must lie on the Simson line BC. Since vertex C is the only point common to AC and BC, we see that Y must be at C, and hence PC is perpendicular to AC. If it exists, therefore, the pole P must lie somewhere on the line perpendicular to AC at C. Similarly, P must lie on the perpendicular to AB at B, and thus P can only be the point where these two lines intersect.

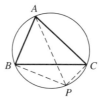

Figure 3.3

But there is another requirement. The pole must also lie on the given circle, and so we need to ask whether or not the intersection point P of the two perpendiculars PB and PC actually lies on the circle, as it appears to do in Figure 3.3. Putting the question another way, we ask if we can find a point P on the circle such that $\angle PBA = 90° = \angle PCA$. The answer should now be obvious. Just choose P so that AP is a diameter. We have shown, therefore, that if we take as a pole a point on the circumcircle of a triangle diametrically opposite a vertex, then the corresponding Simson line is the side opposite that vertex. ∎

In Problem 3.3, we were given a fixed triangle and we were asked to locate a pole for which the corresponding Simson line was parallel to a specific given line: a side of the original triangle. We could ask the same question more generally, as follows. Suppose that we are given some arbitrary line. As the pole P moves around the circle, does the Simson line move in such a way that it can be made parallel to the given line? The answer is yes. In fact, the direction of the Simson line varies in a uniform way as its pole moves: The Simson line turns at exactly half the angular rate at which the pole P moves around the circle. In particular, the Simson line turns through a full 180° as P travels 360° around the circle, and hence it slopes in every possible direction. We can state this a little more precisely, as follows.

(3.4) THEOREM. *Let U and V be points on the circumcircle of $\triangle ABC$. Then the angle between the Simson lines having these points as poles is equal in degrees to half of $\overset{\frown}{UV}$.*

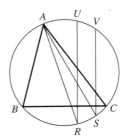

Figure 3.4

There is some ambiguity here since when two lines cross, they determine two angles, and this corresponds to the fact that the points U and V determine two arcs. What is meant by Theorem 3.4 is that the smaller of the two angles is equal in degrees to half of the smaller of the two arcs. We will not give an absolutely rigorous proof that covers all cases of Theorem 3.4, but we hope the argument that follows is convincing.

Proof of Theorem 3.4.   In Figure 3.4, we have extended the perpendiculars from U and V to BC to meet the circle at R and S. By Theorem 3.1, we know that the Simson line having pole U is parallel to AR and the Simson line with pole V is parallel to AS. The angle between these two Simson lines is thus equal to $\angle RAS \overset{\circ}{=} \frac{1}{2} \overset{\frown}{RS}$. Since UR and VS are parallel chords, however, we see that $\overset{\frown}{UV} = \overset{\frown}{RS}$, and the result follows.   ∎

As a special case, we have the following.

(3.5) COROLLARY.   *Two Simson lines for a given triangle are perpendicular if and only if their poles are at opposite ends of a diameter.*   ∎

As we move the pole P around the circumcircle of $\triangle ABC$, we know that the moving Simson line turns so as to slope in every possible direction. We stress, however, that the Simson line does not simply rotate about a single point; its motion is much more complicated. There can be no one point common to all of the Simson lines of $\triangle ABC$ because we have seen that each of the sides of the triangle is a Simson line.

Simson's theorem tells us that if we choose any point on the circumcircle of a triangle and drop perpendiculars from it to the sides of the triangle, extended if necessary, then the feet of these perpendiculars are collinear. This suggests the question of whether or not the same thing can happen if we start with a point not on the circumcircle.

(3.6) THEOREM.   *Given* $\triangle ABC$, *suppose that the feet of the perpendiculars from some point Q to the three sides of the triangle are collinear. Then Q must lie on the circumcircle of* $\triangle ABC$.

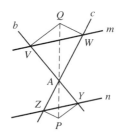

Figure 3.5

In other words, the converse of Simson's theorem is true. To prove it, we need a fairly easy lemma, whose statement, unfortunately, is somewhat complicated. Figure 3.5 should help untangle the hypotheses.

(3.7) LEMMA.   *Suppose that lines m and n are parallel, lines b and c meet at a point A, line m meets b and c at points V and W, respectively, and line n meets b and c at Y and Z, respectively. Perpendiculars to b and c are erected at V and W, and these meet at a point Q. Similarly, the perpendiculars to b and c at Y and Z meet at P. Then points P, A, and Q are collinear.*

Proof.   First, we dispose of some uninteresting "degenerate" cases. If A happens to be one of the points P or Q, there is really nothing to prove. So we can assume that A is neither P nor Q, and it follows that neither of the lines m or n passes through A. Next, we consider what happens if point Q lies on line b. It is not very difficult to see that in this case, line m must be perpendicular to c, and it follows that n is also perpendicular to c, and therefore P must also lie on b. In this case, P, A, and Q are collinear, as desired. We can suppose, therefore, that Q does not lie on b, and similarly, we can assume that Q is not on c and that P is not on either b or c. In particular, Q is different from V and W, and P is different from Y and Z. Next, we observe that V is the foot of the perpendicular from Q to b. Since A lies on b and A is different from V, it follows that AQ is not perpendicular to b. Thus AQ is not parallel to YP, and similarly, AQ is not parallel to ZP.

Figure 3.5 shows one possible configuration: where A lies between lines m and n. It is also possible, of course, that A is above both lines or below both lines. The proof in all cases is identical, however. Our goal is to show that P lies on AQ, as is indicated by the dashed line in the figure.

Since YP is not parallel to AQ, we consider the point R where YP meets AQ, and similarly, we let S be the point where ZP meets AQ. Of course, we expect that R and S are actually the point P, but we do not yet know that P lies on AQ. We propose to show that, in fact, R and S are the same point. Since P is the only point common to lines YP and ZP, it will follow that P, R, and S are all the same point, and in particular, this will show that P lies on AQ, as desired.

Now $YR \| VQ$ since both of these lines are perpendicular to b. It follows using similar triangles that $AR/AQ = AY/AV$, and similarly, we get $AS/AQ = AZ/AW$. But since $YZ \| VW$, we see that $AY/AV = AZ/AW$, and hence we have $AR/AQ = AY/AV = AZ/AW = AS/AQ$. It follows that $AR = AS$, and thus R and S are the same point, as desired. This completes the proof.   ∎

Proof of Theorem 3.6. Let U, V, and W be the feet of the perpendiculars from Q to BC, AC, and AB, respectively, and suppose that these three points all lie on some line m. We have seen that a Simson line can be found parallel to any given line, and so we can choose a pole P (on the circumcircle, of course) for which the corresponding Simson line n is parallel to m. By definition, n runs through the points X, Y, and Z, which are the feet of the perpendiculars from P to lines BC, AC, and AB, respectively. We see now that if we define b to be the line AC and c to be the line AB, we are precisely in the situation of Lemma 3.7.

We conclude from Lemma 3.7 that points P, A, and Q are collinear. Exactly similar reasoning shows that points P, B, and Q are also collinear and that P, C, and Q are collinear too. If P and Q are not the same point, it follows that line PQ runs through all three vertices A, B, and C of $\triangle ABC$. But since these points are the vertices of a triangle, they cannot be collinear, and hence we have a contradiction. Our assumption that P and Q are different must be wrong, therefore, and so we conclude that P and Q are the same point. Since P lies on the circumcircle of $\triangle ABC$, it follows that Q lies on the circumcircle, as required. ∎

There is a really pretty consequence of Simson's theorem (Corollary 3.2) and its converse (Theorem 3.6). This result concerns four lines in **general position**, which means that no two of the lines are parallel and no three of them are concurrent. Note that four lines in general position determine four triangles by taking the lines three at a time, and there are six points of intersection of the lines.

(3.8) COROLLARY. *As illustrated in Figure 3.6, the circumcircles of the four triangles determined by any four lines in general position always go through a common point.*

Proof. Draw two of the circumcircles and observe that one of their points of intersection, which we call P, lies on none of the given lines. Drop perpendiculars from P to each of the four lines, thereby determining four feet, one on each line. By Simson's theorem applied in one of the circles, three of the four feet are collinear, and by a second application of Simson's theorem, in the other circle, another three of the feet are collinear. It follows that all four of the feet of the perpendiculars are collinear, and thus by Theorem 3.6, the point P must lie on all four circumcircles. ∎

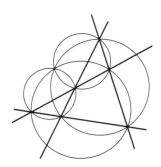

Figure 3.6

There are some intimate connections between the Simson lines associated with a triangle and the nine-point circle of that triangle. Perhaps the most striking of these is the following fact, related to Corollary 3.5. By that result, we know that two Simson lines with poles at the ends of a diameter are perpendicular, and we ask now where these two perpendicular Simson lines meet.

(3.9) THEOREM. *Fix $\triangle ABC$ and let UV be any diameter of its circumcircle. Then the Simson lines of $\triangle ABC$ having poles U and V meet on the nine-point circle of the triangle.*

In fact, if we allow diameter UV to rotate through $180°$, the intersection point of the two Simson lines traces out the whole nine-point circle.

In preparation for proving Theorem 3.9, we recall a useful fact. Given $\triangle ABC$, having orthocenter H, consider the locus of all midpoints of line segments HP joining H to points P on the circumcircle of $\triangle ABC$. By Exercise 1H.10, this locus is a circle, and since the locus clearly contains the three Euler points of the triangle, it must be the unique circle through these points. In other words, the locus of the midpoints of the segments HP is the nine-point circle of $\triangle ABC$.

The following result, which we shall use to prove Theorem 3.9, tells us that every point of the nine-point circle of $\triangle ABC$ lies on an appropriate Simson line.

(3.10) THEOREM. *Let H be the orthocenter of $\triangle ABC$ and let P be any point on the circumcircle of this triangle. Then the midpoint of HP lies on the Simson line of $\triangle ABC$ with pole P.*

Proof.   Figure 3.7 is somewhat complicated, and so we begin with a careful description of how it was drawn. We started with the given $\triangle ABC$ shown in heavy ink and a point P on its circumcircle; we drew altitude AF and marked the orthocenter H on this line, and we drew line segment PH. Next, we drew the line through P perpendicular to side BC and meeting BC at X; we extended PX to meet the circle again at Q, and then we drew AQ.

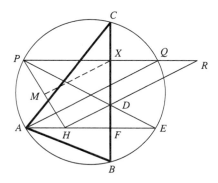

Figure 3.7

The Simson line with pole P certainly goes through X, and we know from Theorem 3.1 that it is parallel to AQ, and so we have drawn (with dashes) the line through X parallel to AQ. This is the Simson line with pole P, and it meets PH at point M. Our task is to show that M is actually the midpoint of PH.

The diagram of Figure 3.7 is completed as follows: Extend altitude AF to meet the circle at E; draw PE meeting side BC at D; and finally, draw HD and extend it to meet the extension of PQ at R. We propose to show that MX is parallel to the base HR of $\triangle PHR$ and that X is the midpoint of side PR of this triangle. It will follow that M is the midpoint of side PH, as required.

Since F is the foot of an altitude of $\triangle ABC$, it lies on the nine-point circle, and hence it is the midpoint of HE since the nine-point circle is the locus of all midpoints of line segments joining H to points on the circumcircle. Thus DF is the perpendicular bisector of HE, and we conclude that $DH = DE$. We now have

$$\angle PQA = \angle PEA = \angle RHE = \angle PRH,$$

where the first equality holds since both angles subtend $\overset{\frown}{PA}$, the second holds by the pons asinorum in isosceles $\triangle DHE$, and the third holds because these are alternate interior angles for parallel lines PR and AE. (Note that PR and AE are parallel because each is perpendicular to BC.) Since $\angle PQA = \angle PRH$, it follows that HR is parallel to AQ, and thus HR is parallel to the Simson line MX. This is one of the two facts we need.

What remains is to prove that X is the midpoint of PR, which we will establish by showing that $\triangle XPD \cong \triangle XRD$. We certainly have $XD = XD$ and $\angle PXD = 90° = \angle RXD$, and so by SAA, it suffices to show that $\angle XPD = \angle XRD$. But $PR \| AE$, and so these angles are equal to $\angle DEH$ and $\angle DHE$, respectively, and these are equal to each other because $\triangle DHE$ is isosceles. The proof is thus complete.   ∎

The proof of Theorem 3.9 is now fairly easy.

Proof of Theorem 3.9.   Let X denote the intersection point of the Simson lines having poles U and V and let H be the orthocenter of the given triangle. By Theorem 3.10, the Simson lines go through the midpoints M and P of HU and HV, respectively, and so the two Simson lines are MX and PX, drawn with heavy ink in Figure 3.8.

We have drawn the nine-point circle of $\triangle ABC$, although the original triangle and its circumcircle are not shown. Midpoints M and P lie on the nine-point circle,

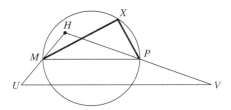

Figure 3.8

and since UV is a diameter of the circumcircle, we know by Corollary 3.5 that $\angle MXP = 90°$. To prove that X lies on the nine-point circle, therefore, it suffices to show that chord MP is actually a diameter of this circle.

Since M and P are the midpoints of two sides of $\triangle UHV$, it follows that $MP = \frac{1}{2}UV$. We recall, however, that for any triangle, the diameter of the nine-point circle is exactly half the diameter of the circumcircle. It follows that chord MP of the nine-point circle has length equal to that of a diameter of this circle, and thus MP must be a diameter, as required. This completes the proof. ∎

There is a remarkable result that we would like to mention before closing this section. Given four points on a circle, we can get four Simson lines by considering each point in turn as a pole and constructing the corresponding Simson line with respect to the triangle formed by the remaining three points. Perhaps the reader can guess what happens.

(3.11) THEOREM.   *The four Simson lines determined by any four distinct points of a circle are concurrent. The point of concurrence, moreover, lies on the nine-point circle of each of the four triangles formed by taking three of the given points.*

We already know that the four nine-point circles go through a common point; that fact was part of Exercise 2C.6. In fact, that exercise actually proves Theorem 3.11. It asserts that the four line segments obtained by joining each of the given points to the orthocenter of the triangle formed by the other three all share a common midpoint. We know that midpoint lies on the four nine-point circles, and by Theorem 3.9, it also lies on the four Simson lines.

We now present a proof of the key step in the solution of Exercise 2C.6.

(3.12) LEMMA.   *Let P, Q, B, and C be four distinct points on a circle and let H and K be the orthocenters of $\triangle PBC$ and $\triangle QBC$, respectively. Then segments PK and QH have a common midpoint.*

Proof.   Refer first to the left diagram of Figure 3.9 and observe that PK and QH are diagonals of quadrilateral $PQKH$. We need to show that these diagonals bisect each

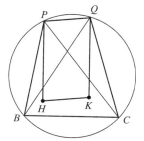
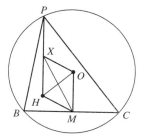

Figure 3.9

other, and so it suffices to show that the quadrilateral is a parallelogram. Certainly, PH and QK are parallel since these lines are altitudes of $\triangle PBC$ and $\triangle QBC$, and hence they are each perpendicular to BC. It suffices, therefore, to show that $PH = QK$.

Let O be the center of the given circle and drop a perpendicular OM from O to BC. (Refer to the right diagram in Figure 3.9 for this.) We will prove that $PH = 2OM$. It follows similarly that $QK = 2OM$, and this shows that $PH = QK$, as required. To prove that $PH = 2OM$, let X be the midpoint of PH and consider quadrilateral $XOMH$. We will show that this is a parallelogram, and it will follow that $OM = XH = \frac{1}{2}PH$, as required.

It suffices to show that diagonals OH and XM bisect each other. The midpoint of OH, we know, is the center of the nine-point circle of $\triangle PBC$. To locate the midpoint of XM, we observe that since O is the center of the given circle, M is the midpoint of chord BC. Also, X is the Euler point opposite M in $\triangle PBC$, and so by Theorem 2.12, segment XM is a diameter of the nine-point circle. Its midpoint is thus also the nine-point center, and so XM and OH have the same midpoint. The proof is now complete. ∎

We mention that there is an easy alternative argument based on Exercise 2C.5. By that exercise, $PH = QK$. Also, PH and QK are both perpendicular to BC, and so they are parallel. Therefore, $PHQK$ is a parallelogram and the result follows.

Exercises 3A

3A.1 Suppose that a Simson line of a triangle goes through its own pole. Show that the pole must be one of the vertices of the triangle.

3A.2 Suppose that a Simson line of a triangle is perpendicular to one of the sides of the triangle. Show that the pole must be one of the vertices of the triangle.

3A.3 Suppose that a Simson line of a triangle goes through the orthocenter of the triangle. Show that the pole must be one of the vertices of the triangle.

3A.4 Show that for an acute angled triangle, every Simson line meets the interior of the triangle. Conversely, show that if every Simson line of a triangle contains a point of the interior, then the angles must all be acute.

3A.5 Prove Theorem 3.1 in the case where the line through P perpendicular to BC is tangent to the circumcircle of $\triangle ABC$. In this case, P and Q coincide, and you must show that PA is parallel to XZ.

3A.6 Given a point P on the circumcircle of $\triangle ABC$, reflect P in each of the sides of the triangle. Show that the three points thus obtained all lie on the line through the orthocenter of $\triangle ABC$ parallel to the Simson line with pole P.

3B   The Butterfly Theorem

The so-called Butterfly theorem is the following pretty result, which is notoriously difficult to prove if you don't know how.

(3.13) THEOREM.   *Suppose that chords PQ and RS of a given circle meet at the midpoint M of chord AB. If X and Y are the points where PS and QR meet AB, respectively, then $XM = YM$.*

We have drawn two possible configurations for this theorem in Figure 3.10, using the "same" chords PQ, RS, and AB in the two diagrams. The only difference is that on the left, points P and R lie on the same side of line AB, while on the right, we have interchanged the labels R and S so that P and R lie on opposite sides of AB. Note that in the latter situation, chord AB had to be extended to meet lines PS and QR. The name "Butterfly" refers to the self-intersecting quadrilateral $PSRQ$, which, with some effort, can be imagined to resemble the eponymous insect.

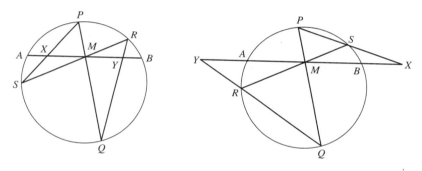

Figure 3.10

We present first an elementary, but somewhat complicated, proof using similar triangles of the case of Theorem 3.13 where the points X and Y lie inside the circle, as in the left diagram of Figure 3.10. Although it is possible to construct a similar proof for the other case, we shall not do so; instead, we present in the next section a much more powerful technique that will allow us to prove both cases of Theorem 3.13 easily.

We begin by extracting the hardest part of the argument as a separate lemma, which is a nice application of similar triangles.

(3.14) LEMMA.   *In Figure 3.11, segments of length x, y, u, v, s, and t are marked. If the angles marked with dots are equal, then*

$$\frac{x^2}{y^2} = \frac{uv}{st}.$$

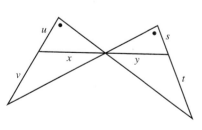

Figure 3.11

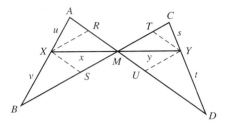

Figure 3.12

Proof.   Label the points in the original diagram as shown in Figure 3.12 and draw XR
and YU parallel to BC and XS and YT parallel to AD.

Since $\triangle XRM \sim \triangle YUM$ and $\triangle XSM \sim YTM$ by AA, we conclude that

$$\frac{RM}{UM} = \frac{x}{y} = \frac{SM}{TM},$$

and this yields

$$\frac{x^2}{y^2} = \frac{RM{\cdot}SM}{TM{\cdot}UM}.$$

Now $\angle A = \angle C$ by hypothesis, and since $\angle AMB = \angle CMD$, we must also
have $\angle B = \angle D$. Also, $\angle AXR = \angle B = \angle D = \angle CYT$, and it follows by AA that
$\triangle AXR \sim \triangle CYT$. We conclude that

$$\frac{u}{s} = \frac{AX}{CY} = \frac{XR}{YT} = \frac{SM}{UM},$$

where the last equality follows because $XRMS$ and $YTMU$ are parallelograms.
Similarly, $v/t = RM/TM$, and we conclude from the equality of the previous
paragraph that $x^2/y^2 = (v/t)(u/s)$, as required.   ∎

Proof of Theorem 3.13.   We prove the case of the theorem that appears in the left
diagram of Figure 3.10. Since $\angle P = \angle R$, we can apply Lemma 3.14 to conclude that

$$\left(\frac{XM}{YM}\right)^2 = \frac{PX{\cdot}XS}{RY{\cdot}YQ}.$$

Now $PX{\cdot}XS = AX{\cdot}XB$ by Theorem 1.35, and similarly, $RY{\cdot}YQ = BY{\cdot}YA$.
Writing $x = XM$, $y = YM$, and $AM = m = BM$, we now see that

$$\frac{x^2}{y^2} = \frac{AX{\cdot}XB}{BY{\cdot}YA} = \frac{(m-x)(m+x)}{(m-y)(m+y)} = \frac{m^2 - x^2}{m^2 - y^2}.$$

Since $m \neq 0$, elementary algebra now yields that $x^2 = y^2$, and we conclude that
$x = y$, as claimed.   ∎

Exercises 3B

3B.1 In Figure 3.13, line segments of length x and y have been drawn from vertex A of $\triangle ABC$ to base BC, and these make equal angles with the sides, as indicated by the dots. These lines divide BC into segments of length u, v, and w, as shown. Find a formula for the ratio x/y in terms of the quantities u, v, and w.

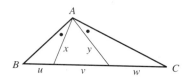

Figure 3.13

3C   Cross Ratios

We now begin a discussion of the theory of cross ratios, which (after all of the preliminary work has been done) can be used to give an easy proof of the Butterfly theorem. As we shall explain presently, the cross ratio of four distinct collinear points is a certain number uniquely determined by these points. To motivate the idea, we consider first the case of two distinct points A and B. These determine the number AB, which is the length of the line segment they determine or the distance between them. Useful and important though it is, the distance function has certain deficiencies. Most obvious among these is the fact that the number AB is not unambiguously determined by the two points; it depends also on the unit of measurement.

Three distinct collinear points A, B, and C determine a number that is independent of the unit of measurement. We can define the associated quantity $r(A, B, C)$ to be the ratio AB/BC. This is a pure (unitless) number that conveys some interesting information. For example, $r(A, B, C) = 1$ if and only if B is the midpoint of AC. Although the ratio $r(A, B, C)$ is independent of the unit of measurement, it suffers from a more subtle deficiency that is also shared by the distance function of two points. To explain this, imagine that A, B, and C are three point sources of light attached to a rigid straight rod. If we observe from afar, we see three apparently collinear lights, but we clearly cannot determine the distances AB and BC without knowing how far away the rod is and how it is oriented with respect to our line of sight. It is also impossible to determine the ratio $r(A, B, C)$ from afar, although in this case, the orientation of the rod is more important than the distance.

Figure 3.14 shows three sets of collinear points $\{A, B, C\}$, $\{X, Y, Z\}$, and $\{P, Q, R\}$ that would look identical to an eye located at point E. In this situation, we say that these three sets are in **perspective** from E. Since we have drawn $AC \parallel PR$, it is an easy exercise using similar triangles to show that $r(A, B, C) = r(P, Q, R)$, but $r(X, Y, Z)$ is definitely unequal to the other two ratios. In the diagram, we have taken Y to be the midpoint of segment XZ, and hence $r(X, Y, Z) = 1$, but it should be clear that B is not the midpoint of AC, and so $r(A, B, C) \neq 1$. Viewing from E, in other words, we

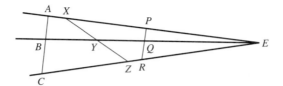

Figure 3.14

cannot determine the ratio $r(A, B, C)$ because we cannot distinguish $\{A, B, C\}$ from $\{X, Y, Z\}$, and yet the associated ratios for these two sets are different.

Remarkably, if we start with four collinear points A, B, C, and D, it is possible to define a unitless quantity, denoted $\mathbf{cr}(A, B, C, D)$, that is invariant under perspective. In other words, if $\{A, B, C, D\}$ and $\{W, X, Y, Z\}$ are sets of distinct collinear points in perspective from a point E, then $\mathbf{cr}(A, B, C, D) = \mathbf{cr}(W, X, Y, Z)$. This means that, in theory, it is possible to "see" the number $\mathbf{cr}(A, B, C, D)$, even if we know neither the distances to the four points nor the orientation in space of the line containing them. We are assuming that the eye E is not collinear with A, B, C, and D; otherwise, it would be impossible even to see that there are four points. As we shall see, it is also necessary for the eye to be able to determine which point is which. We can imagine that the points have four different colors, for instance.

The quantity $\mathbf{cr}(A, B, C, D)$ is called the **cross ratio** of the four distinct collinear points A, B, C, and D, and it is defined by the formula

$$\mathbf{cr}(A, B, C, D) = \frac{AC \cdot BD}{AD \cdot BC}.$$

For example, suppose that the points A, B, C, and D are equally spaced and that they are arrayed in that order along a line. Then $\mathbf{cr}(A, B, C, D) = 4/3$ and $\mathbf{cr}(B, A, C, D) = (BC \cdot AD)/(BD \cdot AC) = 3/4$. The reader should check these calculations.

We were somewhat sloppy when we discussed two *sets* of points as being in perspective from some given point; we should have referred to two *ordered lists* instead. The reason is that the cross ratio of four points depends on the order in which the points occur. For example, we have seen that it is not generally true that $\mathbf{cr}(A, B, C, D) = \mathbf{cr}(B, A, C, D)$. If we expect that, as advertised, the cross ratio will be invariant under perspective from a point, then clearly, the notion of perspective must keep track of the order in which the points occur. A precise definition is as follows. Suppose m and n are lines and P is a point not on either of them. Then points A, B, C, and D of line m are in **perspective from** P with points W, X, Y, and Z of line n, respectively, if W, X, Y, and Z lie on lines PA, PB, PC, and PD, respectively.

Figure 3.15 shows two examples of a perspective from P. In both cases, we started with the same four collinear points A, B, C, and D and the same point P, but we varied the second line n, containing W, X, Y, and Z. Observe that P may lie between two corresponding points, as it lies between A and W in the right diagram, or it may not, as with points Z and D in both diagrams. It is also possible for corresponding points actually to be identical, although we have not drawn a diagram where that occurs.

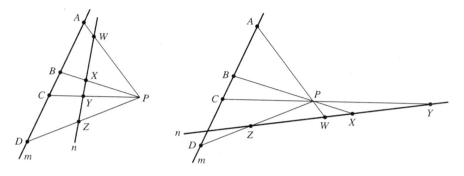

Figure 3.15

(3.15) THEOREM. *Suppose that A, B, C, and D are distinct collinear points in perspective from some point P with distinct collinear points W, X, Y, and Z, respectively. Then* **cr**$(A, B, C, D) =$ **cr**(W, X, Y, Z).

An examination of angles is the key to the proof of Theorem 3.15. Observe that in the left diagram of Figure 3.15, we have

$$\angle APC = \angle WPY \qquad \angle BPC = \angle XPY$$
$$\angle APD = \angle WPZ \qquad \angle BPD = \angle XPZ,$$

but that not all four of these relations hold in the right diagram. In fact, the two equations on the first line are valid in the right diagram because of the equality of vertical angles, but the two equations on the second line do not hold in the right diagram; they must be replaced by

$$\angle APD + \angle WPZ = 180° \qquad \text{and} \qquad \angle BPD + \angle XPZ = 180°.$$

A little experimentation will show that, in general, no matter how the points and lines are arranged, each of the angles $\angle APC$, $\angle BPD$, $\angle APD$, and $\angle BPC$ is either equal to or supplementary to the corresponding angle among $\angle WPY$, $\angle XPZ$, $\angle WPZ$, and $\angle XPY$. Since supplementary angles have equal sines, however, we always have

$$\sin(\angle APC) = \sin(\angle WPY)$$
$$\sin(\angle BPD) = \sin(\angle XPZ)$$
$$\sin(\angle APD) = \sin(\angle WPZ)$$
$$\sin(\angle BPC) = \sin(\angle XPY)$$

whenever collinear points A, B, C, and D are in perspective from P with collinear points W, X, Y, and Z, respectively. To prove Theorem 3.15, therefore, it suffices to show that for any point P not on the line of the collinear points A, B, C, and D, the cross ratio **cr**(A, B, C, D) can be expressed in terms of the four quantities $\sin(\angle APC)$, $\sin(\angle BPD)$, $\sin(\angle APD)$, and $\sin(\angle BPC)$. The theorem is thus an immediate consequence of the following lemma.

(3.16) LEMMA. *Let A, B, C, and D be distinct collinear points and suppose P is any point not on the line through them. Then*

$$\mathbf{cr}(A, B, C, D) = \frac{\sin(\angle APC)\sin(\angle BPD)}{\sin(\angle APD)\sin(\angle BPC)}.$$

Proof. Let h be the perpendicular distance from P to the line containing A, B, C, and D and note that we can compute the area K_{APC} of $\triangle APC$ in two ways. We have $\frac{1}{2}h \cdot AC = K_{APC} = \frac{1}{2}(PA \cdot PC)\sin(\angle APC)$, and there are similar formulas for K_{BPD}, K_{APD}, and K_{BPC}. Thus

$$h \cdot AC = (PA \cdot PC)\sin(\angle APC)$$
$$h \cdot BD = (PB \cdot PD)\sin(\angle BPD)$$
$$h \cdot AD = (PA \cdot PD)\sin(\angle APD)$$
$$h \cdot BC = (PB \cdot PC)\sin(\angle BPC),$$

and we see that if we divide the product of the first two of these equations by the product of the second two, the lengths PA, PB, PC, PD, and h all cancel and we get

$$\mathbf{cr}(A, B, C, D) = \frac{AC \cdot BD}{AD \cdot BC} = \frac{\sin(\angle APC)\sin(\angle BPD)}{\sin(\angle APD)\sin(\angle BPC)},$$

as required.   ∎

The following shows how Theorem 3.15 can be used in certain types of problems.

(3.17) PROBLEM. Given that $AB = 3$ and $BC = 1$ in Figure 3.16, find CD.

Solution. There hardly seems to be enough information here. In fact, the location of point P is completely arbitrary and cannot be determined from the given data, but nevertheless, we shall see that it is possible to determine CD, as required. Before we explain how to do this, however, we would like to describe the procedure that was used to draw the diagram of Figure 3.16. We started with collinear points A, B, and C with distances as given, and we chose point P not on line AC but otherwise completely arbitrarily. Lines PA, PB, and PC were drawn and point Q, different

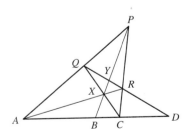

Figure 3.16

from A, and P was chosen arbitrarily on AP. Next, CQ was drawn, meeting PB at a point labeled X, and then AX was drawn, meeting PC at R. Finally, point D was determined as the point where QR meets the original line AC. Remarkably, the information contained in this paragraph is sufficient to compute the cross ratio $\mathbf{cr}(A, B, C, D)$, even without specifying the distances AB and BC. We state this result as a separate lemma.   ■

(3.18) LEMMA. *In the configuration of points and lines shown in Figure 3.16, we have* $\mathbf{cr}(A, B, C, D) = 2$.

Proof.   Since collinear points Q, Y, R, and D are in perspective from point P with collinear points A, B, C, and D, we know from Theorem 3.15 that $\mathbf{cr}(Q, Y, R, D) = \mathbf{cr}(A, B, C, D)$. On the other hand, if we consider the perspective from X instead of from P, we see that Q, Y, R, and D are in perspective from C, B, A, and D, respectively, and thus $\mathbf{cr}(Q, Y, R, D) = \mathbf{cr}(C, B, A, D)$. It follows, therefore, that

$$\frac{AC \cdot BD}{AD \cdot BC} = \mathbf{cr}(A, B, C, D) = \mathbf{cr}(C, B, A, D) = \frac{CA \cdot BD}{CD \cdot BA}.$$

Since the numerators of these fractions are equal, the denominators must be equal too, and so if we write $x = AB$, $y = BC$, and $z = CD$, we have $(x + y + z)y = AD \cdot BC = CD \cdot BA = zx$. Thus $y(y + z) = xz - xy$, and we can recompute the numerator

$$AC \cdot BD = (x + y)(y + z) = x(y + z) + y(y + z) = x(y + z) + x(z - y) = 2xz$$

Since the denominator is equal to xz, we have $\mathbf{cr}(A, B, C, D) = 2$, as claimed.   ■

Solution to Problem 3.17, continued.   We have $AB = 3$ and $BC = 1$, and we write $CD = z$, an unknown. By Lemma 3.18, we have $2 = \mathbf{cr}(A, B, C, D) = 4(1 + z)/(4 + z)$, and solving this, we get $z = 2$.   ■

More generally, if we follow the notation of the proof of Lemma 3.18 and write $AB = x$, $BC = y$, and $CD = z$ in the situation of Figure 3.16, we can solve for z in terms of x and y to get $z = y(x + y)/(x - y)$. (Check that the substitution of $x = 3$ and $y = 1$ yields $z = 2$, as we saw before.)

Now imagine distorting Figure 3.16, keeping all lines straight, so that B moves toward A, decreasing x and increasing y, while holding $x + y$ constant. It is clear from both the diagram and the formula that z increases as B moves. It "blows up" to infinity when $x = y$, which occurs when the moving point reaches the midpoint of AC. What does it mean for point D to move "infinitely far" to the right? What is this trying to tell us? Clearly, the conclusion we should draw here is that QR is parallel to AC when AB is a median of $\triangle APC$. This is indeed a theorem, although the argument of this paragraph is questionable as a proof. In fact, there is a way to interpret parallel lines as intersecting "at infinity" and thereby to deal with infinite quantities in cross ratios. The branch of geometry where this is done is called projective geometry, but we will not pursue that any further here.

We mention one other case where the formula $z = y(x + y)/(x - y)$ gives a nonsensical result. If $y > x$, then according to the formula, $z = CD$ is negative. Obviously, the length of a line segment cannot be negative, so what is going on here? The answer is that if $BC = y > x = AB$, then the diagram of Figure 3.16 is not correct. In that case, the point D, where QR meets AC, lies to the left of A and not to the right of C, as shown. Interestingly, the distance CD as given by our formula is as correct as it can be. The actual distance is the absolute value of the computed negative quantity, and we can interpret the negative sign as meaning that D lies that many units on the wrong side of C, the side opposite from that indicated in the picture.

To apply the theory of cross ratios to prove the Butterfly theorem, we need to define the cross ratio of four distinct points on a circle. Just as points on a line are said to be collinear, we shall say that points on a circle are **cocircular**. Note that if three or more distinct points are cocircular, then there is a unique circle containing all of them.

Suppose A, B, C, and D are any four distinct cocircular points. We define the **cross ratio** of these points to be the quantity

$$\mathbf{cr}(A, B, C, D) = \frac{\sin\left(\frac{1}{2}\,\widehat{AC}\right)\sin\left(\frac{1}{2}\,\widehat{BD}\right)}{\sin\left(\frac{1}{2}\,\widehat{AD}\right)\sin\left(\frac{1}{2}\,\widehat{BC}\right)},$$

where the arcs, of course, are on the common circle through the four given points. There are several potential ambiguities that we need to resolve before we can say that this really makes sense. First, observe that we can measure the arcs in either degrees or radians, as we please, provided that we interpret the sine function accordingly. The "standard" function $f(x) = \sin(x)$ assumes that x is measured in radians, but if we wish, we can measure angles and arcs in degrees and use the function $g(x) = \sin(x°)$. Observe, for example, that $f(\pi/2) = 1 = g(90)$. Next, note that it is not actually true that two points X and Y of a given circle define a unique arc; they actually determine two arcs whose angular measures sum to $360°$. Nevertheless, the quantity $\sin(\frac{1}{2}\,\widehat{XY})$ is unambiguously defined because the two possible meanings of $\frac{1}{2}\,\widehat{XY}$ correspond to two angles that sum to $180°$, and thus they have equal sines.

By the discussion of the previous paragraph, the quantity $\mathbf{cr}(A, B, C, D)$ is unambiguously defined for any four distinct cocircular points A, B, C, and D. For example, the reader can check that if $ABCD$ is a square, then $\mathbf{cr}(A, B, C, D) = 2$ and $\mathbf{cr}(B, A, C, D) = 1/2$. There remains a potential ambiguity in our notation, however: The definition of $\mathbf{cr}(A, B, C, D)$ is different if the points A, B, C, and D are collinear rather than cocircular. We are safe, however, because four distinct points can never be both collinear and cocircular, and hence at most one of the two definitions of $\mathbf{cr}(A, B, C, D)$ applies. There should thus be no danger of confusion.

We are using the same name "cross ratio" and the same notation "$\mathbf{cr}(\ ,\ ,\ ,\ )$" for four collinear points and for four cocircular points because there is an intimate connection between these two concepts, as the following theorem demonstrates.

(3.19) THEOREM.   *Let A, B, C, and D be four distinct cocircular points and suppose P is a point on the same circle, different from all of them. Given a line not through P, let W, X, Y, and Z be the four necessarily distinct and collinear points where*

PA, PB, PC, and PD, respectively, meet the given line. Then $\mathbf{cr}(A, B, C, D) = \mathbf{cr}(W, X, Y, Z)$.

Proof.   Figure 3.17 shows one of the many possible configurations for this theorem. We know from Lemma 3.16 that

$$\mathbf{cr}(W, X, Y, Z) = \frac{\sin(\angle WPY)\sin(\angle XPZ)}{\sin(\angle WPZ)\sin(\angle XPY)}$$

and by definition,

$$\mathbf{cr}(A, B, C, D) = \frac{\sin\left(\frac{1}{2}\,\widehat{AC}\right)\sin\left(\frac{1}{2}\,\widehat{BD}\right)}{\sin\left(\frac{1}{2}\,\widehat{AD}\right)\sin\left(\frac{1}{2}\,\widehat{BC}\right)} .$$

It suffices, therefore, to show that the four sines in the first formula are equal, respectively, to the four sines in the second formula.

In Figure 3.17, we see that $\angle WPY = \angle APC \overset{\circ}{=} \frac{1}{2}\,\widehat{ABC}$. Thus $\sin(\angle WPY) = \sin(\frac{1}{2}\,\widehat{AC})$, where in this equation of sines, we need not specify which of the two possible arcs $\widehat{AC}$ we mean. Similarly, $\angle XPY = \angle BPC \overset{\circ}{=} \frac{1}{2}\,\widehat{BC}$, where we refer here to the arc between B and C that excludes point P. Thus $\sin(\angle XPY) = \sin(\frac{1}{2}\,\widehat{BC})$, where again, because sines of supplementary angles are equal, we can afford to be sloppy in our designation of the arc.

In the configuration shown in Figure 3.17, the other two equalities of sines for the angles that involve point Z hold for a slightly different reason. We see that $\angle XPZ$ is supplementary to $\angle BPD \overset{\circ}{=} \frac{1}{2}\,\widehat{BCD}$, and thus $\angle XPZ \overset{\circ}{=} \frac{1}{2}\,\widehat{BPD}$, and we have $\sin(\angle XPZ) = \sin(\frac{1}{2}\,\widehat{BD})$. Similarly, $\angle WPZ \overset{\circ}{=} \frac{1}{2}\,\widehat{APD}$, and thus $\sin(\angle WPZ) = \sin(\frac{1}{2}\,\widehat{AD})$, as required. This proves the equality of the two cross ratios in the diagram of Figure 3.17.

In general, it is easy to see that each angle is always equal in degrees to half of one of the two possible arcs corresponding to it. The precise rule is that we need to use the arc that contains P when P lies between exactly one of the two pairs of corresponding points, and otherwise, we use the arc excluding P. In Figure 3.17, for instance, P lies between D and Z, but it does not lie between A and W, B, and X, or C and Y. This is the reason we had to choose the arcs containing P for the angles involving Z in the previous paragraph. The four sines of angles, therefore, are always equal to the four sines of arcs, and the proof is complete. ∎

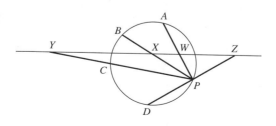

Figure 3.17

In the situation of Theorem 3.19, we shall refer to P as the **projection point**. This result is very powerful, and it is especially useful in the restricted case where the line WZ cuts the circle at two of the original four points. To demonstrate this, we consider the following problem, which appears to be very difficult by other methods.

(3.20) PROBLEM.   A point on the circumcircle of a square is joined to the two most distant vertices, thereby cutting the nearest side of the square into three pieces. Find the length of the middle segment if the other two pieces have lengths 3 and 10.

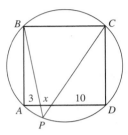

Figure 3.18

Solution.   In Figure 3.18, denote the intersection points of PB and PC with AD by R and S, respectively, and write $RS = x$. By Theorem 3.19, with projection point P, we have

$$2 = \mathbf{cr}(A, B, C, D) = \mathbf{cr}(A, R, S, D) = \frac{AS \cdot RD}{RS \cdot AD} = \frac{(x+3)(x+10)}{x(x+13)},$$

where the first equality holds because $ABCD$ is a square.

Elementary algebra now yields the equation $x^2 + 13x - 30 = 0$, and this quadratic equation has roots $x = 2$ and $x = -15$. Since we are looking for a positive number, we reject $x = -15$ and conclude that the length of the middle segment is 2. ∎

We now give the promised quick proof of the Butterfly theorem using cross ratios. We repeat Figure 3.10 here as Figure 3.19 and we recall that M is the midpoint of AB and that our task is to prove that $MX = MY$ in each of the two diagrams.

Proof of Theorem 3.13.   In either diagram, we see that by using P as the projection point in Theorem 3.19, we get $\mathbf{cr}(A, X, M, B) = \mathbf{cr}(A, S, Q, B)$. With R as the projection point, however, we get $\mathbf{cr}(A, M, Y, B) = \mathbf{cr}(A, S, Q, B)$, and so we deduce that $\mathbf{cr}(A, X, M, B) = \mathbf{cr}(A, M, Y, B)$.

Now write $x = MX$, $y = MY$, and $AM = m = MB$. Working in the left diagram of Figure 3.19, we see that

$$\mathbf{cr}(A, X, M, B) = \frac{AM \cdot XB}{AB \cdot XM} = \frac{m(x+m)}{2mx} = \frac{1}{2} + \frac{m}{2x}$$

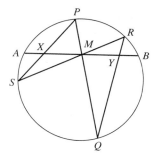 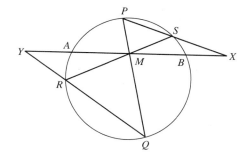

Figure 3.19

and

$$\mathbf{cr}(A, M, Y, B) = \frac{AY \cdot MB}{AB \cdot MY} = \frac{(y + m)m}{2my} = \frac{1}{2} + \frac{m}{2y}.$$

Since we know that these are equal, we deduce that $x = y$, as required.

In the right diagram, a similar calculation yields that

$$\mathbf{cr}(A, X, M, B) = \frac{m(x - m)}{2mx} = \frac{1}{2} - \frac{m}{2x}$$

and

$$\mathbf{cr}(A, M, Y, B) = \frac{(y - m)m}{2my} = \frac{1}{2} - \frac{m}{2y}$$

and we deduce that $x = y$ in this case too.   ∎

This proof of the Butterfly theorem yields some additional information, which we discuss somewhat informally. Suppose PQ, RS, and AB are any three chords of a circle that are concurrent at a point M that is not necessarily the midpoint of AB. Let X and Y, respectively, be the intersections of AB with PS and with QR, as in Figure 3.19. We now have five distinct collinear points A, X, M, Y, and B, and the first part of our proof of the Butterfly theorem shows that $\mathbf{cr}(A, X, M, B) = \mathbf{cr}(A, M, Y, B)$. We stress that this part of the argument did not rely on the assumption that $AM = MB$ in the Butterfly theorem. We shall say that five distinct collinear points A, X, M, Y, and B have the **butterfly property** whenever $\mathbf{cr}(A, X, M, B) = \mathbf{cr}(A, M, Y, B)$, and we observe that this property is invariant under perspective since the cross ratios are invariant by Theorem 3.15.

We claim now that if we had started with three concurrent chords of an ellipse instead of a circle, the five points A, X, M, Y, and B would nevertheless continue to enjoy the butterfly property. To see why this is so, we use the fact that an ellipse is a conic section. Without going into detail here, we can exploit this by imagining the following situation.

We are in a room with a point source of light in the ceiling and a clean white floor. We have a planar sheet of glass on which an ellipse has been drawn with opaque ink, and we hold it so that a shadow of the ellipse is cast onto the floor. The fact that the ellipse is a conic section tells us that it is possible to position and tilt the glass in such a way that the shadow is a circle. In fact, this can be done so that the center of the circle is directly below the light, but we shall not need this additional information. While our assistant holds the glass steady, with the ellipse casting a circular shadow, we draw three concurrent chords on the ellipse, and we construct and mark the points $\mathbf{A}$, $\mathbf{X}$, $\mathbf{M}$, $\mathbf{Y}$, and $\mathbf{B}$, as we discussed previously. The boldface font indicates that we used opaque ink, and it allows us to distinguish these five points from their shadows A, X, M, Y, and B, which are points on the floor.

Since shadows of lines are lines, we see that the shadows of the three concurrent chords of the ellipse are three concurrent chords of its circular shadow. Furthermore, the shadow of the butterfly pattern on the ellipse is a butterfly pattern on the circle, as in Figure 3.19, and we deduce that the five collinear shadow points A, X, M, Y, and B satisfy the butterfly property. The key observation here is that the five shadow points A, X, M, Y, and B are in perspective from the light L with the five points $\mathbf{A}$, $\mathbf{X}$, $\mathbf{M}$, $\mathbf{Y}$, and $\mathbf{B}$ marked on the glass. It follows that the latter five points also enjoy the butterfly property, as claimed.

Suppose now that the point of concurrence $\mathbf{M}$ of the three chords of the ellipse happens to be the midpoint of chord $\mathbf{AB}$. (Caution: It does not follow that M is the midpoint of AB.) Since we know that $\mathbf{cr}(A, X, M, B) = \mathbf{cr}(A, M, Y, B)$, we can now apply the second part of the cross-ratio proof of the Butterfly theorem to deduce that $\mathbf{XM} = \mathbf{MY}$. This part of the argument did not rely on the assumption that we started with a circle. The preceding discussion shows that Theorem 3.13, the Butterfly theorem, holds for ellipses as well as for circles.

Exercises 3C

3C.1 Suppose that A, B, X, and Y are collinear and distinct. If $AX/AY = BX/BY$, show that exactly one of the points A and B lies on line segment XY.

3C.2 Suppose that A, B, and C are distinct and collinear and that X and Y lie on the line through them.

 a. If $\mathbf{cr}(A, B, C, X) = \mathbf{cr}(A, B, C, Y)$, show that either X and Y are the same point, or else exactly one of A and B lies on segment XY.
 b. If $\mathbf{cr}(A, B, C, X) = \mathbf{cr}(A, B, C, Y)$ and also $\mathbf{cr}(C, B, A, X) = \mathbf{cr}(C, B, A, Y)$, show that X and Y must be the same point.

 NOTE: This provides a tool that can be used to prove that points X and Y on the line through A, B, and C are actually identical.

3C.1 Points W and X are chosen on side AB of $\triangle ABC$ and points Y and Z are chosen on side AC. Suppose that $\mathbf{cr}(A, W, X, B) = \mathbf{cr}(A, Y, Z, C)$ and that $WY \| XZ$. Prove that $XZ \| BC$.

 HINT: Let T be the point where the parallel to XZ through B meets line AC. Note that neither A nor Y can lie on segment TC and use Exercise 3C.2(a) to show that T is C.

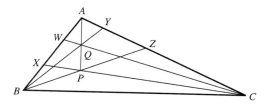

Figure 3.20

3C.2 In Figure 3.20, points P and Q were chosen in the interior of $\triangle ABC$ and collinear with A. Lines BQ and BP meet side AC at Y and Z, and lines CQ and CP meet side AB at W and X. Show that $\mathbf{cr}(A, W, X, B) = \mathbf{cr}(A, Y, Z, C)$.

3C.3 In the situation of Figure 3.20, suppose that lines WY and XZ meet at a point R. Show that R lies on line BC.

HINT: Let T be the point where line RB crosses line AC. Use Exercise 3C.2(b) to show that C and T are the same point.

3C.4 In Figure 3.21, we have chosen a point X on the extension of diagonal BD of square $ABCD$, and we dropped a perpendicular CP from C to AX. If CP meets DB at Y and $BY = XD$, show that $BY = BA$.

HINT: Points A, B, C, D, and P are cocircular.

3C.5 Figure 3.22 shows a self-intersecting hexagon $ABCDEF$ inscribed in a circle. Sides AB and DE intersect at U, sides BC and EF intersect at V, and sides CD and FA intersect at W. Show that points U, V, and W are collinear.

HINT: Let X and Y be the points where FA and CD, respectively, meet UV. Prove that X and Y are the same point by showing that $\mathbf{cr}(P, T, Q, X) = \mathbf{cr}(P, U, Q, V) = \mathbf{cr}(P, T, Q, Y)$, where P and Q are the points where UV meet the circle, as shown, and T is the unlabeled point where AD meets UV. Since points P and Q can be interchanged, the same reasoning yields that $\mathbf{cr}(Q, T, P, X) = \mathbf{cr}(Q, T, P, Y)$.

NOTE: Even if hexagon $ABCDEF$ inscribed in a circle is not self-intersecting or if different pairs of sides intersect, we can almost always define three points U, V, and W as the intersection points of the three pairs of lines AB and DE, BC

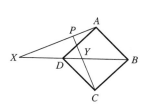

Figure 3.21

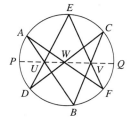

Figure 3.22

and EF, and CD and FA. We can run into difficulty because the lines in one or more of these pairs might be parallel. Also, we note that the sides of the hexagon may need to be extended to construct U, V, and W. It is a theorem of B. Pascal (1623–1662) that points U, V, and W are always collinear. In fact, this works for hexagons inscribed in arbitrary conic sections and not just in circles.

3C.6 Let A, B, C, and D be distinct cocircular points and fix a line m not through A. Let X, Y, and Z be the points where AB, AC, and AD, respectively, meet line m and let W be the intersection point of m with the tangent at A to the circle through A, B, C, and D. Show that $\mathbf{cr}(A, B, C, D) = \mathbf{cr}(W, X, Y, Z)$.

NOTE: This result can be viewed as the limiting case of Theorem 3.19 as the point P of that theorem approaches A.

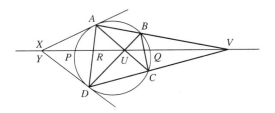

Figure 3.23

3C.7 Let $ABCD$ be a quadrilateral inscribed in a circle, as shown in Figure 3.23. If U is the intersection of its diagonals and V is the intersection of sides AB and CD, show that the tangents to the circle at A and at D meet line UV at the same point. Assume that neither of these tangents is parallel to UV.

HINT: Let X and Y be the points where the two tangents meet UV and show that $\mathbf{cr}(P, U, Q, X) = \mathbf{cr}(P, V, Q, R) = \mathbf{cr}(P, U, Q, Y)$, where P and Q are the points where line UV crosses the circle and R is the intersection of side AD with UV, as in the diagram. This calculation requires four applications of Exercise 3C.8.

NOTE: Similar reasoning shows that the tangents at B and D also intersect at a point on line UV.

3C.8 Let AB be a diameter of a circle and suppose that UV is a perpendicular chord. Show that $\mathbf{cr}(A, U, B, V) = 2$.

3C.9 In the situation of the previous problem, fix an arbitrary point P on line AB and construct the point Q on AB as follows. Draw line PU and let X be the point other than U where PU meets the circle. If PU is tangent to the circle at U, take X to be U. Now let Q be the point where VX meets AB. Show that point Q is independent of the choice of the chord UV perpendicular to AB. Figure 3.24 shows one possible configuration for this problem.

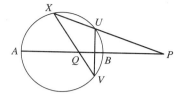

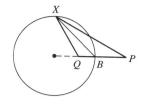

Figure 3.24

Figure 3.25

3C.10 In Figure 3.25, we are given $\triangle QXP$ with angle bisector XB, where X and B lie on a circle whose center lies on line QP. Show that if Y is any point of the circle other than B, then YB bisects $\angle QYP$.

3D   The Radical Axis

Consider three circles, each externally tangent to the other two. In Problem 2.28, we showed that in this situation, the three common tangent lines are concurrent. The phrase "common tangent line" is somewhat imprecise. Here and in what follows, we intend this to refer to the common tangent at the point of tangency of the two circles. Actually, as indicated in Figure 3.26, something much more general is going on here.

First, as the diagram on the left demonstrates, the requirement that the circles be *externally* tangent in Problem 2.28 is not really necessary. In fact, *tangency* is really irrelevant too. To explain this, we observe that there is an interesting line naturally associated with any pair of circles having at least one point in common. For tangent circles, this line is the common tangent; for nontangent intersecting circles, the line to which we refer is the line through the two points of intersection, which we somewhat imprecisely call the common secant. (Note that the common tangent is really just a limiting case of this common secant. If we gradually increase the separation between the centers of the two circles, the two intersection points coalesce to a single point of tangency, and the common secant becomes the common tangent.)

If we start with any three circles, each having a point in common with the other two, we get three interesting lines, one from each pair of circles. We will prove the remarkable fact that, as shown in Figure 3.26, these three lines are always concurrent.

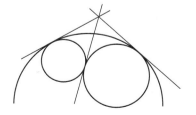

Figure 3.26

Actually, there is an exception. If the three centers are collinear, then the three lines will be parallel.

(3.21) THEOREM.   *Given three circles with noncollinear centers, assume that every two of them have a point in common. If for each pair of circles, we draw as appropriate either the common secant or the common tangent, then the three lines thus constructed are concurrent.*

Consider what happens if we hold two of the circles fixed and vary the third. Of course, the line determined by the two fixed circles remains fixed, and the other two lines will move. We observe, however, that the (moving) point of intersection of the two moving lines is constrained by Theorem 3.21 to travel along the third (fixed) line. To see that there is still more going on here, consider what happens if we try the same experiment in a situation where the two fixed circles have no point in common, but the varying circle meets both of them. An example of this appears in Figure 3.27, where we show four positions and sizes for the variable circle. To make the diagram appear less cluttered, we chose one of the two fixed circles inside the other, but this arrangement is not essential. The only requirement is that the two fixed circles should not be concentric.

We again have two moving lines, and again we observe the behavior of their intersection point. Remarkably, its trajectory still appears to be a straight line. In the figure, P, Q, R, and S are collinear.

What is the locus of the intersection point of the two moving common secants? It is clearly some object determined by the two fixed circles, and in the case where these circles have a point in common, the locus is their common secant or tangent line. We will prove that the locus is always a line, even when the two given circles have no point in common. This line, which is dashed in Figure 3.27, is called the radical axis of the two circles, but before we can define this term officially, we need to introduce an auxiliary idea.

Given a point P and a circle of radius r centered at some point O, we say that the **power** of P with respect to the given circle is the quantity $p = d^2 - r^2$, where $d = PO$ is the distance from the point to the center of the circle. The points exterior to the circle clearly have positive power. The interior points have negative power, and the points of the circle itself have power $p = 0$.

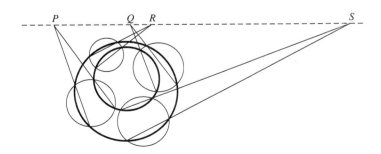

Figure 3.27

To make the definition of the power of a point with respect to a circle seem a little less arbitrary, we recall Theorem 1.35. In that result, we showed that if P is any point not on a given circle and we select any line through P meeting the circle at two points X and Y, then the quantity $PX \cdot PY$ is a constant, independent of the choice of the line. The following lemma tells us that this mysterious constant is either the power of the point P or its negative. We shall not actually use most of this lemma; its purpose here is to relate the power of a point to ideas that we have seen previously.

(3.22) LEMMA.   *Fix a circle and a point P and let p be the power of P with respect to the given circle.*
   a. *If P lies outside the circle and a line through P cuts the circle at X and Y, then $PX \cdot PY = p$.*
   b. *If P is inside the circle on chord XY, then $PX \cdot PY = -p$.*
   c. *If P lies on the line tangent to the circle at point T, then $(PT)^2 = p$.*

Proof.   Assume first that P lies outside the circle. By Theorem 1.35, the quantity $PX \cdot PY$ will be the same for all lines through P that meet the circle in two points. It is no loss, therefore, to assume that the line XY through P actually goes through the center of the circle, which we denote O. Thus XY is a diameter, and we can assume that X is the nearer of these two points to P. Writing $PO = d$ and $XO = r$, the radius, we see that $PX = PO - XO = d - r$ and $PY = PO + YO = d + r$. Thus $PX \cdot PY = (d - r)(d + r) = d^2 - r^2 = p$, as required

Suppose now that P lies inside the circle. By Theorem 1.35, the quantity $PX \cdot PY$ is a constant, independent of the particular chord XY through P. To prove (b), therefore, we can assume that chord XY is a diameter, and we can further assume that P lies on the segment OX. We see that $PX = XO - PO = r - d$ and $PY = YO + PO = r + d$. Thus $PX \cdot PY = (r - d)(r + d) = r^2 - d^2 = -p$, as required.

Finally, if PT is tangent to the circle at T, we can use (a) and a limit argument to show that $(PT)^2 = p$. Alternatively, we observe that $\triangle OTP$ is a right triangle with side $OT = r$ and hypotenuse $PT = d$. It follows by the Pythagorean theorem that $(PT)^2 = d^2 - r^2 = p$, as required.   ∎

(3.23) THEOREM.   *Fix two circles, centered at distinct points A and B. Then there exist points whose powers with respect to the two given circles are equal. The locus of all such points is a line perpendicular to AB.*

Proof.   It is convenient to use coordinate geometry for this proof. We can suppose that points A and B lie on the x axis so that A is the point $(a, 0)$ and B is $(b, 0)$, where $a \neq b$. If P is an arbitrary point with coordinates (x, y), then $(PA)^2 = y^2 + (x - a)^2$ and $(PB)^2 = y^2 + (x - b)^2$.

Writing r and s to denote the radii of the given circles centered at A and B, respectively, we see that the powers of P with respect to the two circles are equal if and only if

$$y^2 + (x - a)^2 - r^2 = y^2 + (x - b)^2 - s^2 .$$

The y^2 terms cancel, and when we expand the parentheses, the x^2 terms cancel too. The condition that the two powers of P are equal thus reduces to the linear equation $a^2 - 2ax - r^2 = b^2 - 2bx - s^2$. Since $b - a$ is nonzero, this is equivalent to

$$x = \frac{r^2 - s^2 + b^2 - a^2}{2(b - a)},$$

where we observe that the right side is some constant. Since this is the equation of a line perpendicular to the x axis, the proof is complete. ∎

Finally, we can define the object in the heading of this section. Given two circles having different centers, their **radical axis** is the line consisting of all points that have equal powers with respect to the two circles.

(3.24) COROLLARY.   *If two circles intersect at two points A and B, then their radical axis is their common secant AB. If two circles are tangent at a point T, then their radical axis is their common tangent at T.*

Proof.   A point common to two circles has power 0 with respect to each of them, and thus its two powers are equal and the point lies on the radical axis. If A and B are two different points common to two circles, then A and B both lie on the radical axis, which we know is a line. It follows that the radical axis is the line AB.

In the case where two circles are tangent at T, then since T is on both circles, it lies on the radical axis. To see that the radical axis is tangent to each circle at T, it suffices to show that T is the only point where this line meets either circle. This is clear, however, because if a point P of the radical axis lies on one of the circles, its power with respect to that circle, and hence with respect to the other circle too, is 0. It follows that P lies on both circles, and hence P is the unique point common to the two circles, namely, T. ∎

It is now clear that the following easy result includes Theorem 3.21.

(3.25) COROLLARY.   *Given three circles with noncollinear centers, the three radical axes of the circles taken in pairs are distinct concurrent lines.*

Proof.   Since the radical axis of a pair of circles is perpendicular to the line of centers of the circles, it follows from the noncollinearity of the three centers that the three radical axes are distinct and nonparallel. Every two of them, therefore, have a point of intersection.

For any point P, let us write p_1, p_2, and p_3 to denote the powers of P with respect to the three given circles. For points on one radical axis, we have $p_1 = p_2$, and on another, we have $p_2 = p_3$. At the point P, where these two radical axes meet, we have $p_1 = p_2 = p_3$, and thus $p_1 = p_3$, and P also lies on the third radical axis. ∎

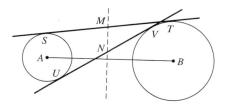

Figure 3.28

We mention that in the situation of Corollary 3.25, the unique point common to the three radical axes is called the **radical center** of the three circles.

Given two nonconcentric circles printed on a piece of paper, how can we actually find and draw their radical axis? If the two circles intersect at two points, this is an easy task: Just draw the line through the two points. If the two circles are tangent, the radical axis is the common tangent line, and this is only a bit more difficult to draw.

But how can we draw the radical axis if the two circles have no points in common? If the circles are external to each other, we can draw a line tangent to both, and we let S and T be the two points of tangency, as shown in Figure 3.28.

If M is any point on line ST, then by Lemma 3.22(c), we know that distances MS and MT are the powers of point M with respect to the two circles. It follows that if we take M to be the midpoint of segment ST, as in Figure 3.26, then $MS = MT$, and so the two powers are equal. The midpoint M thus lies on the radical axis, and since we know that the radical axis is perpendicular to the line of centers, it suffices to draw the perpendicular to this line through M to complete the construction. Alternatively, we can draw one of the other three lines tangent to both circles, and we let U and V be its two points of tangency. The midpoint N of segment UV must also lie on the radical axis, which can thus be constructed by drawing line MN. We can avoid the possibility that M and N are the same point if we choose the second tangent appropriately.

If one of the two circles is inside the other, there are no lines tangent to both circles, and the method of the previous paragraph cannot be used. Although it may not be obvious how to proceed, it is really not hard to construct the radical axis in this situation too. The method we are about to describe is suggested by Figure 3.27, but it actually works in all cases. In fact, it is probably even easier to carry out than the construction of the previous paragraph.

Draw an auxiliary circle meeting each of the two given circles in two points and draw the line through each of these pairs of points. These two lines are the radical axes of the auxiliary circle with each of the two original circles, and we know by Corollary 3.25 that the point P where these lines meet must lie on the radical axis that we seek. Now choose a second auxiliary circle and perform a similar construction to obtain a point Q. Since both P and Q are known to lie on the radical axis of the two given circles, we can complete our construction by drawing line PQ.

Given any two fixed nonconcentric circles, we mentioned previously that their radical axis is the locus of all points P that can be obtained using an auxiliary circle, as in the previous paragraph. We have already argued that every point constructed this way must lie on the radical axis, but to prove that this is actually the locus, we must also show that every point on this line can be obtained from some choice of an auxiliary circle. Given any point P on the radical axis, draw two lines through P, one for each of the given circles, intersecting it in two points. This gives four points, two on each circle, and we see that a circle through these four points would serve as an auxiliary circle that yields P. Our task, then, is to show that these four points actually are cocircular.

(3.26) COROLLARY.   *Given points A and B on one circle and C and D on another, let P be the intersection of lines AB and CD. Then P lies on the radical axis of the two given circles if and only if the four points A, B, C, and D are cocircular.*

Proof.   If the four points all lie on some circle, then by Corollary 3.24, lines AB and CD are the radical axes of this circle with each of the given circles. By Corollary 3.25, their intersection P lies on the radical axis of the two given circles.

Conversely, suppose P lies on the radical axis of the two given circles and name these circles **X** and **Y**, where A and B lie on **X** and C and D lie on **Y**. Consider the circle **Z** through A, B, and C and observe that AB is the radical axis of **X** and **Z**. Since P lies on this radical axis and also on the radical axis of the original two circles **X** and **Y**, it follows by Corollary 3.25 that P also lies on the radical axis of **Y** and **Z**. But C also lies on this radical axis, and we conclude that the radical axis of circles **Y** and **Z** must be the line PC. This line goes through D, however, and thus D has equal powers with respect to circles **Y** and **Z**. Since D lies on **Y**, we conclude that it also lies on **Z**, and the proof is complete.   ∎

Exercises 3D

3D.1 Consider the three circles whose diameters are the sides of a given triangle. Show that the radical center of these circles is the orthocenter of the triangle.

3D.2 In Figure 3.29, the common chord PQ of two circles bisects line segment AB, where A and B lie on the circles as shown. If X and Y are the other points where AB meets the two circles, show that $BX = AY$.

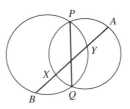

Figure 3.29

CHAPTER FOUR

Ceva's Theorem and Its Relatives

4A Ceva's Theorem

We know several theorems that have the form "the three 'somethings' of a triangle are concurrent." In every triangle, for example, the three medians are concurrent, and so too are the three angle bisectors and the three altitudes. Notice that in each of these situations, the three concurrent somethings are three lines, each of them passing through one vertex of the triangle. Of course, there are many other ways to define three lines passing through the vertices of a triangle, and it seems amazing how often there is a theorem that guarantees that these three lines must go through a common point.

But not every concurrence theorem associated with a triangle has the form we are discussing here. We know, for example, that the perpendicular bisectors of the sides of a triangle are always concurrent, but this is not a result of the type we are considering because the perpendicular bisectors generally do not go through the vertices. A concurrence fact that actually is an example of the phenomenon we have in mind, however, is the following.

(4.1) PROBLEM. Let P, Q, and R be the points of tangency of the incircle of $\triangle ABC$ with sides BC, CA, and AB, respectively, as in Figure 4.1. Show that lines AP, BQ, and CR are concurrent.

The point of concurrency of the three lines in Problem 4.1 is sometimes called the **Gergonne point** of the triangle. As is apparent in Figure 4.1, the Gergonne point need

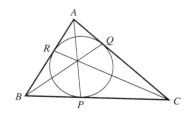

Figure 4.1

not be the center of the circle, and thus in general, the three lines are not the angle bisectors. We shall present the solution of Problem 4.1 as an application of the theorem of Giovanni Ceva (1648–1737) that is the main result of this section.

(4.2) THEOREM (Ceva). *Let AP, BQ, and CR be three lines joining the vertices of $\triangle ABC$ to points P, Q, and R on the opposite sides. Then these three lines are concurrent if and only if*

$$\frac{AR}{RB}\frac{BP}{PC}\frac{CQ}{QA} = 1.$$

We shall refer to any line going through exactly one vertex of a triangle as a **Cevian** of the triangle. By Theorem 4.2, we see that if we wish to prove that three Cevians are concurrent (as in Problem 4.1, for example), all we need to do is compute the product of the three fractions in Ceva's theorem and show that the resulting quantity is equal to 1. In general, we will refer to this quantity as the **Cevian product** associated with the three given Cevians. Of course, we can also think of this product of three fractions as the quotient obtained by dividing the product of three segment lengths by the product of another three segment lengths.

One way to remember the definition of the Cevian product is to imagine traveling around the triangle from A through R to B, then from B through P to C, and then along the third side from C through Q and back to A. For each side of the triangle, we get one fraction: first AR/RB, then BP/PC, and finally CQ/QA. The Cevian product is just the product of these three fractions.

Solution to Problem 4.1.   Recall that in Figure 4.1, we know that $AR = QA$, $BP = RB$, and $CQ = PC$. (Actually, by Lemma 2.27, we know more: The quantities AR, BP, and CQ are equal to $s - a$, $s - b$, and $s - c$, respectively, in the usual notation. We shall not need that additional information, however.) Observe that AR occurs in the numerator of the Cevian product and that QA occurs in the denominator. These two equal quantities thus cancel when we compute the product, and similarly, BP cancels with RB and CQ cancels with PC. All six quantities cancel, therefore, and the value of the Cevian product is 1. It follows by Ceva's theorem that the three Cevians are concurrent, as required. ∎

Notice that Ceva's theorem is an if and only if statement. Thus whenever we have three concurrent line segments joining the vertices of a triangle with points on the opposite sides, the Cevian product must be trivial. By trivial, we mean, of course, that the product is equal to 1. Suppose, for example, that our three Cevians are the medians of $\triangle ABC$ so that P, Q, and R are the midpoints of BC, CA, and AB, respectively, and thus $AR = RB$, $BP = PC$, and $CQ = QA$. Since the medians are concurrent, we know that the Cevian product has to be trivial, and indeed it is because here too the six quantities cancel. Next, consider the case where the Cevians AP, BQ, and CR are the angle bisectors. Again, the Cevian product must be trivial since the angle bisectors are always concurrent, and we can confirm that by recalling Theorem 1.12, which tells us that an angle bisector of a triangle divides the opposite side into pieces whose lengths are proportional to the nearer sides of the triangle. Using the usual notation, where we write a, b, and c to denote the lengths of the sides opposite vertices A, B, and C, we have $AR/RB = b/a$, $BP/PC = c/b$, and $CQ/QA = a/c$. The Cevian product $(AR/RB)(BP/PC)(CQ/QA)$ is thus equal to $(b/a)(c/b)(a/c)$, and again everything cancels and the Cevian product is trivial, as expected.

We need a simple lemma for the proof of Theorem 4.2, and we state it somewhat more generally than is really necessary.

(4.3) LEMMA. *Given distinct points A and B and a positive number μ, there is exactly one point X on the line segment AB such that $AX/XB = \mu$. Also, there is at most one other point on the line AB for which this equation holds.*

Proof. View X as a variable point and let $f(X)$ be the function whose value at X is the quantity AX/XB. Thus $f(X)$ is a nonnegative real number, and it is defined everywhere except when $X = B$. As X moves from A toward B along segment AB, we see that AX increases and XB decreases, and thus $f(X)$ is monotonically increasing from 0 when X is at A, and it approaches infinity as X approaches B. There is thus exactly one point X between A and B where $f(X) = \mu$.

If X is on line AB outside of segment AB, there are just two possibilities: Either B is between X and A, or else A is between X and B. In the first case, $AX = XB + BA$ and $f(X) = AX/XB = 1 + (BA/XB) > 1$. Otherwise, $AX = XB - BA$ and $f(X) = AX/XB = 1 - (BA/XB) < 1$. For any given value $f(X) = \mu$, therefore, at most one of these two situations can occur depending on whether $\mu > 1$ or $\mu < 1$.

If B is between X and A, the function $f(X) = 1 + (BA/XB)$ is monotonically decreasing as X moves farther from B. Otherwise, $f(X) = 1 - (BA/XB)$ is monotonically increasing as X gets farther from B. In either case, we see that there can be at most one point X such that $f(X) = \mu$. ∎

We shall also need an elementary algebraic fact about ratios of real numbers. If two ratios are equal (say, $a/b = c/d$), then we automatically get two more ratios equal to these two, namely, $(a+c)/(b+d)$ and $(a-c)/(b-d)$. (Of course, we need to assume that $b + d$ is nonzero for the first of these and that $b - d$ is nonzero for the second.) To see why this works, write $\lambda = a/b = c/d$ so that $a = \lambda b$ and $c = \lambda d$. Then $a+c = \lambda(b+d)$ and $a-c = \lambda(b-d)$, and thus $(a+c)/(b+d) = \lambda = (a-c)/(b-d)$, as we wanted. We will refer to these as the addition and subtraction principles for ratios.

We are now ready to present the surprisingly easy proof of Ceva's powerful theorem.

Proof of Theorem 4.2.   Assume first that the three Cevians are concurrent at some point T, as in Figure 4.2. We compute the Cevian product by considering the areas of certain triangles. View BP and PC as the bases of $\triangle ABP$ and $\triangle APC$, respectively, and observe that these triangles have equal heights. It follows that BP/PC is the ratio of the areas of these two triangles. Since segments BP and PC can also be viewed as the bases of $\triangle TBP$ and $\triangle TPC$, and these two triangles also have equal heights, we conclude that

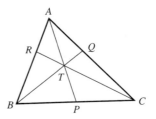

Figure 4.2

$$\frac{K_{ABP}}{K_{APC}} = \frac{BP}{PC} = \frac{K_{TBP}}{K_{TPC}},$$

where as usual, we are writing K_{XYZ} to denote the area of an arbitrary $\triangle XYZ$. By the subtraction principle for ratios, we deduce that

$$\frac{BP}{PC} = \frac{K_{ABP} - K_{TBP}}{K_{APC} - K_{TPC}} = \frac{K_{ABT}}{K_{CAT}}.$$

Exactly similar reasoning yields

$$\frac{AR}{RB} = \frac{K_{CAT}}{K_{BCT}} \qquad \text{and} \qquad \frac{CQ}{QA} = \frac{K_{BCT}}{K_{ABT}},$$

and thus we can compute the Cevian product. We have

$$\frac{AR}{RB}\frac{BP}{PC}\frac{CQ}{QA} = \frac{K_{CAT}}{K_{BCT}}\frac{K_{ABT}}{K_{CAT}}\frac{K_{BCT}}{K_{ABT}} = 1,$$

as we wanted.

To prove the converse, we assume that the Cevian product is trivial. We prove that AP, BQ, and CR are concurrent by defining T to be the intersection of AP and BQ in Figure 4.2 and showing that line CR must also pass through T. This is equivalent, of course, to showing that line CT goes through R, and so we let R' be the point where line CT actually does meet side AB. Then CR' is a Cevian that is concurrent with AP and BQ, and so by the first part of the proof, the corresponding Cevian product is trivial.

We now have

$$\frac{AR'}{R'B}\frac{BP}{PC}\frac{CQ}{QA} = 1 = \frac{AR}{RB}\frac{BP}{PC}\frac{CQ}{QA},$$

where the second equality is by assumption. Cancellation yields that $AR'/R'B = AR/RB$, and we write μ to denote the common value of these two equal ratios. By Lemma 4.3, there can only be one point on line segment AB for which $AX/XB = \mu$. But R and R' both lie on segment AB, and for each of these points, we get the same ratio μ. Thus R and R' must actually be the same point, and hence CT goes through R, as required.   ∎

We can also think about Ceva's theorem from the point of view of centers of mass in physics. Imagine that $\triangle ABC$ is manufactured from weightless rods, and we place nonzero masses m_A, m_B, and m_C at vertices A, B, and C, respectively. The center of mass of side BC will be at that point P on side BC where the moments $m_B BP$ and $m_C PC$ are equal, and thus $BP/PC = m_C/m_B$. We know that in every balancing experiment, side BC behaves as though its total mass were concentrated at the point P, and thus we can pretend that the entire mass $m_B + m_C$ of side BC is placed at point P. The center of mass T of the entire triangle, therefore, must lie on the Cevian AP. By exactly similar reasoning, we know that T also lies on Cevians BQ and CR, where Q and R are the centers of masses of sides CA and AB, respectively. The three Cevians determined by the masses m_A, m_B, and m_C are thus guaranteed to be concurrent at the center of mass T of the triangle. As a check, we compute the Cevian product. We saw that $BP/PC = m_C/m_B$, and similar reasoning shows that $CQ/QA = m_A/m_Q$ and $AR/RB = m_B/m_A$. We therefore have

$$\frac{AR}{RB}\frac{BP}{PC}\frac{CQ}{QA} = \frac{m_B}{m_A}\frac{m_C}{m_B}\frac{m_A}{m_C} = 1 \, ,$$

and the Cevian product is trivial, as expected.

Now, suppose that AP, BQ, and CR are Cevians of some triangle $\triangle ABC$ and assume that the corresponding Cevian product is trivial. We will show that it is always possible to choose nonzero masses m_A, m_B, and m_C so that if we place these masses at vertices A, B, and C, respectively, the centers of mass of the three sides will be at P, Q, and R. We are continuing to assume, of course, that the actual sides are weightless and that all of the mass is concentrated at the vertices. Assuming that we can find such masses, this will provide an alternative proof that the three given Cevians must be concurrent. This is because each Cevian will have to pass through the center of mass of the triangle.

To find appropriate nonzero masses m_A, m_B, and m_C, we view these quantities as unknowns and we attempt to solve the three simultaneous equations $m_B BP = m_C PC$, $m_C CQ = m_A QA$, and $m_A AR = m_B RB$. One solution for this system of equations would be to set each of the unknowns to zero, but this is of no value to us since centers of mass of massless objects are undefined. We can choose our units of mass so that $m_A = 1$, and this forces $m_C = QA/CQ$ and $m_B = AR/RB$ from the second and third equations. To verify that the first equation also holds for these values of m_A, m_B, and m_C, we need to check that the quantities

$$m_B BP = \frac{AR \cdot BP}{RB} \qquad \text{and} \qquad m_C PC = \frac{QA \cdot PC}{CQ}$$

are equal. But $AR \cdot BP \cdot CQ = QA \cdot RB \cdot PC$ since we are assuming that the Cevian product is equal to 1, and thus $m_B BP = m_C PC$, as desired. This completes the "physics" proof that three Cevians must be concurrent when the corresponding Cevian product is trivial.

As a lagniappe, this method of mass points enables us to compute the position along each Cevian of the point of concurrence of three concurrent Cevians. We know, for example, that the point of concurrence of the three medians of a triangle lies two thirds of the way along each median, and we would like to have analogous information for any three concurrent Cevians.

We have seen that the center of mass T of $\triangle ABC$ with masses m_A, m_B, and m_C at the vertices lies on Cevian AP. Furthermore, we can assume that the whole mass $m_B + m_C$ of side BC is concentrated at P. It follows that $m_A AT = (m_B + m_C)TP$. Using the fact that $AP = AT + TP$, some elementary algebra yields

$$\frac{AT}{AP} = \frac{m_B + m_C}{m_A + m_B + m_C}.$$

We can now plug in $m_A = 1$, $m_B = AR/RB$, and $m_C = QA/CQ$, and with a little algebra, we deduce the following formula, which we state as a theorem.

(4.4) THEOREM.   *Let AP, BQ, and CR be Cevians in $\triangle ABC$, where P, Q, and $R lie on sides BC, CA, and AB, respectively. If these Cevians are concurrent at a point T, then*

$$\frac{AT}{AP} = \frac{AR \cdot CQ + QA \cdot RB}{AR \cdot CQ + QA \cdot RB + RB \cdot CQ},$$

and similar formulas hold for BT/BQ and CT/CR. ■

Theorem 4.4 tells us where to look along Cevian AP to find the point T of concurrence. More specifically, it tells us what fraction of the route from A to P we must traverse to find T. We can check this somewhat unpleasant formula in the case where we are dealing with three medians and we already know that the intersection, the centroid of $\triangle ABC$, lies two thirds of the way along each median. In this situation, we can write $AR = r = RB$ and $CQ = q = QA$. Plugging this into the formula of Theorem 4.4, we get $AT/AP = (2rq/3rq) = 2/3$, as expected.

Exercises 4A _____

4A.1 Let T be the Gergonne point of $\triangle ABC$. Recall that this is the point of concurrence of the Cevians in the situation of Problem 4.1. Show that if T coincides with the incenter or the circumcenter or the orthocenter or the centroid of $\triangle ABC$, then the triangle must be equilateral.

4A.2 Let U and V be points on sides AB and AC, respectively, of $\triangle ABC$ and suppose that UV is parallel to BC. Show that the intersection of UC and VB lies on median AM.

    NOTE: This is essentially Problem 1.30, but you should now be able to find a much easier proof than we gave in Chapter 1.

4A.3 Show that the lines joining the vertices of a triangle to the points of tangency of the opposite exscribed circles are concurrent.

4A.4 In Figure 4.2, if $AR/AB = 2/3$ and $BP/BC = 2/3$, compute CQ/CA and AT/AP.

4A.5 Given three concurrent Cevians in a triangle, show that the three lines obtained by joining the midpoints of the Cevians to the midpoints of the corresponding sides are concurrent.

HINT: Consider the medial triangle.

4A.6 Suppose Cevians AP, BQ, and CR are concurrent at point T, as in Figure 4.2, so that $\triangle ABC$ is decomposed into six small triangles. If the areas of $\triangle ART$, $\triangle BPT$, and $\triangle CQT$ are equal, show that in fact all six small triangles have equal areas.

HINT: To make the algebra a little neater, assume that units have been chosen so that the areas of $\triangle ART$, $\triangle BPT$, and $\triangle CQT$ are each equal to 1.

4B   Interior and Exterior Cevians

In the statement and proof of Theorem 4.2 and in all of the examples we have considered so far, Cevians AP, BQ, and CR "begin" at vertices of $\triangle ABC$, they cut across the interior of the triangle, and they "terminate" at points lying on the opposite sides of the triangle. We refer to such lines as **interior** Cevians. Interestingly, Ceva's theorem almost remains valid even if we expand our definition and allow **exterior** Cevians. These are lines that join a vertex of a triangle to a point on an *extension* of the opposite side, and which thus do not cut across the interior of the triangle. We shall insist, however, that a Cevian must go through just one vertex of the triangle. Otherwise, the Cevian product would be the meaningless expression $0/0$. For example, AP is an exterior Cevian if point P lies on line BC, but it does not lie on the line *segment* BC, and in particular, P is not one of the points B or C.

We shall see that if Cevians AP, BQ, and CR are concurrent, then the Cevian product is guaranteed to equal 1, even if not all of the given Cevians are interior. If it is suitably interpreted and modified, the converse statement also remains true when exterior Cevians are allowed. If we know that the Cevian product is trivial, we shall see that under appropriate additional conditions, we can conclude that the Cevians are concurrent. To understand the additional complications in the statement and proof of this part of the theorem, consider Figure 4.3, which shows the three possible configurations that we need to consider.

In each of these three diagrams, AP, BQ, and CR are Cevians that are concurrent at a point T. On the left, all are interior Cevians, but in the other two diagrams, AP is interior while both BQ and CR are exterior. A little experimentation shows that these are indeed the only possibilities, except for a possible renaming of points, and we observe that if three Cevians are concurrent, then the number of interior Cevians among them is necessarily either one or three. In short, given three concurrent Cevians, the number of interior Cevians among them must be odd. Although it is possible for the Cevian product to be trivial when the number of interior Cevians is even, we see that in this case, the Cevians cannot be concurrent. It follows that some additional assumption must be made

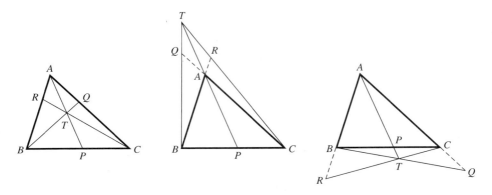

Figure 4.3

if we wish to conclude that three Cevians are concurrent when the corresponding Cevian product is trivial. The appropriate additional assumption is precisely that the number of interior Cevians should be odd.

There is yet another complication. Consider, for example, the middle diagram in Figure 4.3 and imagine point T moving upward along line AP, farther and farther from A. As T moves, we adjust Q so that it remains at the intersection of lines BT and AC, and similarly, we keep R at the intersection of lines CT and AB. At every stage, AP, BQ, and CR are concurrent Cevians with a Cevian product equal to 1. In the limiting situation, therefore, the Cevian product will still be 1, but the three Cevians will be parallel, and not concurrent. It follows that in the general form of Ceva's theorem, the word *concurrent* must be replaced by the phrase *concurrent or parallel*.

Sometimes it is helpful to pretend that parallel lines meet at some imaginary point that is "infinitely far" away. With that fiction, we could say that three parallel Cevians are concurrent, and we would avoid having to consider an exceptional case in Ceva's theorem. To be a little more precise, we imagine that each line in the plane contains one extra ideal point (or point at infinity) in addition to its real points. We want *every* two lines, parallel or not, to have exactly one point in common, and so we declare that parallel lines share a common ideal point, but nonparallel lines have distinct ideal points. Also, since we want every two points to determine a line, we create one ideal line consisting of all the ideal points. All of this can be made precise and rigorous, and what results is an extension of the Euclidean plane in which every two lines have a common point and every two points lie on a common line. This extended plane, which is called the **projective plane**, is often useful for avoiding annoying special cases and exceptions.

We can expand our notion of Cevians a bit further. By our definition, almost every line through a vertex of a triangle is a Cevian of that triangle. In fact, there are exactly three exceptions for each vertex. Through vertex A of $\triangle ABC$, for example, the exceptions are lines AB and AC and the line through A parallel to BC. We have explicitly excluded AB and AC from consideration, and the third exclusion is a consequence of our requirement that a Cevian through A must be a line of the form AP, where P is some point of line BC other than B or C. Of course, the line parallel to BC through A contains no such point P, and so by our current definition, this line would not qualify as a Cevian.

If we work in the projective plane, then parallel lines meet at some ideal point at infinity. In this situation, the line through A parallel to BC is a Cevian, and the corresponding point P is the ideal point on line BC. We adopt this point of view now and declare that the line through A parallel to BC is a Cevian. But there is a problem when we try to extend Ceva's theorem in this situation. If P is the ideal point on BC, how should we compute the factor BP/PC in the Cevian product? The answer is easy: We simply define $BP/PC = 1$ in this case so that the Cevian product becomes $(AR/RB)(CQ/QA)$. This is reasonable since the limit of BP/PC is 1 as P moves off to infinity.

We can now state the fully general form of Ceva's theorem.

(4.5) THEOREM. *Let AP, BQ, and CR be Cevians of $\triangle ABC$, where points P, Q, and R lie on lines BC, CA, and AB, respectively. Then these Cevians are either concurrent or parallel if and only if an odd number of them are interior and*

$$\frac{AR}{RB}\frac{BP}{PC}\frac{CQ}{QA} = 1.$$

Proof. We will only sketch a proof here, omitting many of the details. If the three Cevians are concurrent or parallel, we observe from an appropriate diagram that an odd number of them must be interior, and in particular, at least one is interior, and so we can assume that AP is interior. If the three lines are actually concurrent and not parallel, we can assume that we are in the situation of one of the three diagrams of Figure 4.3, and we can compute the Cevian product using areas of appropriate triangles, just as we did in the proof of Theorem 4.2.

In all cases, we have

$$\frac{AR}{RB} = \frac{K_{ACR}}{K_{RCB}} = \frac{K_{ATR}}{K_{RTB}}$$

$$\frac{BP}{PC} = \frac{K_{BAP}}{K_{PAC}} = \frac{K_{BTP}}{K_{PTC}}$$

$$\frac{CR}{RA} = \frac{K_{CBQ}}{K_{QBA}} = \frac{K_{CTQ}}{K_{QTA}}.$$

Next, we apply the addition and subtraction principles for ratios, as we did in the proof of Theorem 4.2. (The appropriate principle in each case depends on which of the three diagrams of Figure 4.3 is under consideration.) This yields

$$\frac{AR}{RB} = \frac{K_{CAT}}{K_{BCT}}, \qquad \frac{BP}{PC} = \frac{K_{ABT}}{K_{CAT}}, \qquad \text{and} \qquad \frac{CR}{RA} = \frac{K_{BCT}}{K_{ABT}},$$

and thus

$$\frac{AR}{RB}\frac{BP}{PC}\frac{CQ}{QA} = 1,$$

as required. In the case where the three Cevians are parallel, we get the same conclusion either by taking limits or by a direct argument using similar triangles. (See Problem 4B.2.)

To prove the converse, assume that the number of interior Cevians is odd and that the Cevian product is trivial. There is nothing to prove if the three lines are parallel, and so we can assume that two of them (say, AP and BQ) meet at some point T, and we use essentially the same argument as in the proof of Theorem 4.2 to show that line CR must also pass through T. We propose to show that line CT goes through R, and so we let R' be the point where CT meets line AB, and we work to show that R and R' are the same point. Now CR' is a Cevian that is concurrent with AP and BQ, and so we know that the corresponding Cevian product is trivial. Reasoning exactly as in the proof of Theorem 4.2, we deduce that $AR'/R'B = AR/RB$.

Next, we observe that either both of the Cevians CR and CR' are interior or else neither is. This is because the total number of interior Cevians in each of the sets $\{AP, BQ, CR\}$ and $\{AP, BQ, CR'\}$ is odd. The first set contains an odd number of interior Cevians by hypothesis, and the second set contains an odd number since these three Cevians are known to be concurrent by construction. It follows either that both of the points R and R' lie on the line segment AB or else neither of them does. The fact that R and R' must be the same point is now a consequence of Lemma 4.3. ∎

(4.6) PROBLEM.   Use Ceva's theorem to find an alternative proof of the fact that the altitudes of a triangle are concurrent. (See Theorem 2.10.)

Solution.   Observe that there are exactly three possibilities for our given $\triangle ABC$. Either all of the angles of the triangle are acute, as in the left diagram of Figure 4.4, one of the angles (say, $\angle A$) is obtuse, as in the right diagram of the figure, or the triangle has a right angle. In the latter situation, where we have a right triangle, each altitude clearly goes through the right angle, and so there is really nothing to prove in this case. We can thus assume that one of the two diagrams of Figure 4.4 applies, and thus the altitudes AP, BQ, and CR are Cevians that we want to show are concurrent. We see that in either case, the number of altitudes that are interior is odd, and hence it suffices by Theorem 4.5 to show that the Cevian product $(AR/RB)(BP/PC)(CQ/QA)$ is trivial.

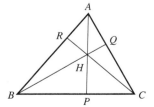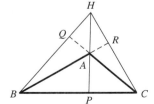

Figure 4.4

In the left diagram of Figure 4.4, we have

$$AR = b\cos(A) \qquad RB = a\cos(B)$$
$$BP = c\cos(B) \qquad PC = b\cos(C)$$
$$CQ = a\cos(C) \qquad QA = c\cos(A),$$

where, as usual, we have written a, b, and c to denote the lengths of the sides opposite A, B, and C, respectively. Each of a, b, and c thus occurs once in the numerator and once in the denominator of the Cevian product, and similarly, each of $\cos(A)$, $\cos(B)$, and $\cos(C)$ occurs once in the numerator and once in the denominator. Everything cancels, therefore, and the Cevian product is trivial, as required.

In the right diagram, where $\angle BAC > 90°$, exactly the same equations hold for the lengths of the six line segments provided that we interpret $\angle A$ as referring to the *exterior* angle of the triangle at A so that $\cos(A)$ will be positive. It follows that in this case too the Cevian product is trivial. ∎

Exercises 4B

4B.1 In Figure 4.5, compute each of the ratios XP/PY and BQ/QC in terms of the lengths $r = AX$, $s = XB$, $t = AY$, and $u = YC$. Show that the only way these ratios can be equal is if both ratios equal 1, and show that in that case, XY is parallel to BC.

HINT: There are two ways to see concurrent Cevians in Figure 4.5. There are three concurrent interior Cevians for $\triangle ABC$, but there are also three concurrent Cevians for $\triangle AXY$.

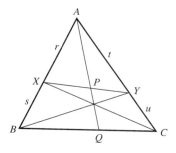

Figure 4.5

4B.2 Show using similar triangles that if Cevians AP, BQ, and CR are parallel, then the Cevian product is trivial. (See Figure 4.6.)

HINT: Let a, b, and c denote the lengths of the sides of $\triangle ABC$ and write $x = BP$ and $y = PC$. Express each of the six lengths that appear in the Cevian product in terms of the five quantities a, b, c, x, and y.

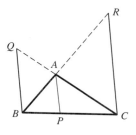

Figure 4.6

4B.3 Consider $\triangle ABC$ in Figure 4.7. Cevians CR and BQ are concurrent with line AT, which is parallel to BC. If we view AT as a Cevian of $\triangle ABC$ in the generalized sense, then we know that the corresponding Cevian product is $(AR/RB)(CQ/QA)$. Show without reference to Ceva's theorem that this quantity is equal to 1.

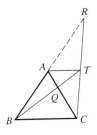

Figure 4.7

4C   Ceva's Theorem and Angles

Suppose that AP is a Cevian in $\triangle ABC$, where as usual, P lies on line BC. This Cevian determines two distances: BP and PC, neither of which is zero since P is not allowed to be one of the points B or C. Given three Cevians, one through each vertex of $\triangle ABC$, then, of course, six distances are determined. If we know these six distances, then by Ceva's theorem, we can decide whether or not the three given Cevians are concurrent.

In addition to determining the two distances BP and PC, the Cevian AP also determines two angles: $\angle BAP$ and $\angle PAC$, neither of which is zero. Given three Cevians, therefore, one through each vertex of $\triangle ABC$, there are six angles determined. For some problems, it would be useful to be able to determine from a knowledge of these six angles whether or not the three Cevians are concurrent. It should be clear that it is impossible to determine the six distances appearing in the Cevian product from a knowledge of these six angles, but nevertheless, and perhaps surprisingly, if we know the six angles, it is not hard to compute the value of the Cevian product. The six angles can thus be used to determine whether or not the three Cevians are concurrent, and as we shall see, this is exactly what we need to solve a number of interesting problems.

In fact, the Cevian product is always equal to the quantity we call the **angular Cevian product**, which is the product of the three fractions in the statement of the following result.

(4.7) THEOREM.   *Suppose that AP, BQ, and CR are Cevians in $\triangle ABC$. Then the corresponding Cevian product is equal to*

$$\frac{\sin(\angle ACR)}{\sin(\angle RCB)} \frac{\sin(\angle BAP)}{\sin(\angle PAC)} \frac{\sin(\angle CBQ)}{\sin(\angle QBA)}.$$

In particular, the three Cevians are concurrent or parallel if and only if an odd number of them are interior and this angular Cevian product is equal to 1.

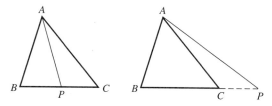

Figure 4.8

Proof.   The three factors of the Cevian product are, of course, AR/RB, BP/PC, and CQ/QA, and we will use the law of sines to express each of these in terms of angles. (To review the law of sines, see the discussion in Section 1E or refer to Theorem 2.3.) We work first with the ratio BP/PC, and we begin with the case where the Cevian AP is interior, as in the left diagram of Figure 4.8.

Working in $\triangle ABP$, we have

$$\frac{BP}{\sin(\angle BAP)} = \frac{AP}{\sin(\angle B)},$$

and similarly, in $\triangle ACP$, we have

$$\frac{PC}{\sin(\angle PAC)} = \frac{AP}{\sin(\angle C)}.$$

If we solve these equations for BP and PC and then divide and cancel AP, we obtain

$$\frac{BP}{PC} = \frac{\sin(\angle BAP)}{\sin(\angle PAC)}\frac{\sin(\angle C)}{\sin(\angle B)}.$$

In the case where AP is an exterior Cevian, it turns out that we get exactly the same formula. To see this, consider the right diagram of Figure 4.8 and notice that there is no change at all in the formula that we obtained from the law of sines in $\triangle ABP$. In $\triangle PAC$, however, we observe that the angle opposite side AP is not the original $\angle C = \angle ACB$, but instead, it is the corresponding exterior angle of the triangle, namely, $\angle ACP = 180° - \angle C$. But the sine of any angle is equal to the sine of its supplement, and so when we apply the law of sines, we get the same formula we had previously.

In all cases, therefore, we have

$$\frac{BP}{PC} = \frac{\sin(\angle BAP)}{\sin(\angle PAC)}\frac{\sin(\angle C)}{\sin(\angle B)},$$

and similarly, we get

$$\frac{CQ}{QA} = \frac{\sin(\angle CBQ)}{\sin(\angle QBA)}\frac{\sin(\angle A)}{\sin(\angle C)} \quad \text{and} \quad \frac{AR}{RB} = \frac{\sin(\angle ACR)}{\sin(\angle RCB)}\frac{\sin(\angle B)}{\sin(\angle A)}.$$

When we multiply the three ratios to compute the Cevian product, we get

$$\frac{AR}{RB}\,\frac{BP}{PC}\,\frac{CQ}{QA} = \frac{\sin(\angle ACR)}{\sin(\angle RCB)}\,\frac{\sin(\angle BAP)}{\sin(\angle PAC)}\,\frac{\sin(\angle CBQ)}{\sin(\angle QBA)}\,,$$

as required.   ∎

If the three Cevians happen to be the angle bisectors of the triangle, it is obvious that the angular Cevian product is trivial, and this is consistent with the fact that the angle bisectors are always concurrent. It is a little more difficult, however, to see that the ordinary Cevian product is trivial in this case, although we have done that calculation. On the other hand, if the Cevians are the three medians, it is immediate that the ordinary Cevian product is trivial, but this is not so obvious for the angular Cevian product. Although the two forms of the Cevian product always have equal values, we see that they are often not equally easy to evaluate. For any particular problem, therefore, one or the other of these products might be the more appropriate tool.

An application of the angular Cevian product is the following surprising result.

(4.8) THEOREM.   *Given an arbitrary* $\triangle ABC$, *build three outward-pointing triangles* BCU, CAV, *and* ABW, *each sharing a side with the original triangle, as illustrated in Figure 4.9. Assume that*

$$\angle BAW = \angle CAV,$$

$$\angle CBU = \angle ABW, \quad and$$

$$\angle ACV = \angle BCU.$$

Assume further that lines AU, BV, *and* CW *cut across the interior of* $\triangle ABC$, *as in the figure. Then lines* AU, BV, *and* CW *are concurrent.*

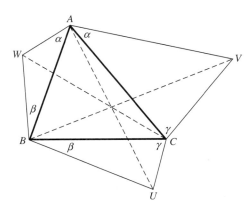

Figure 4.9

Proof. Lines AU, BV, and CW are interior Cevians, and so by Theorem 4.7, it suffices to show that the angular Cevian product is trivial. In other words, we must establish that

$$\frac{\sin(\angle ACW)}{\sin(\angle WCB)}\frac{\sin(\angle BAU)}{\sin(\angle UAC)}\frac{\sin(\angle CBV)}{\sin(\angle VBA)} = 1.$$

For clarity of notation, we write $\alpha = \angle BAW = \angle CAV$, and similarly, we let β and γ denote the measures of the other two pairs of equal angles, as indicated in Figure 4.9.

To compute $\sin(\angle ACW)$, we use the law of sines in $\triangle ACW$ to deduce that

$$\frac{AW}{\sin(\angle ACW)} = \frac{CW}{\sin(\angle WAC)}.$$

Since $\angle WAC = \angle A + \alpha$, we have

$$\sin(\angle ACW) = \frac{AW\,\sin(\angle A + \alpha)}{CW},$$

and similarly, interchanging the roles of A and B, we see that

$$\sin(\angle WCB) = \frac{BW\,\sin(\angle B + \beta)}{CW},$$

and we obtain

$$\frac{\sin(\angle ACW)}{\sin(\angle WCB)} = \frac{AW\,\sin(\angle A + \alpha)}{BW\,\sin(\angle B + \beta)}.$$

The ratio AW/BW can be computed using the law of sines again, this time in $\triangle ABW$. We have

$$\frac{AW}{\sin(\beta)} = \frac{BW}{\sin(\alpha)},$$

and thus

$$\frac{AW}{BW} = \frac{\sin(\beta)}{\sin(\alpha)}.$$

Substitution of this into our previous formula yields

$$\frac{\sin(\angle ACW)}{\sin(\angle WCB)} = \frac{\sin(\beta)\,\sin(\angle A + \alpha)}{\sin(\alpha)\,\sin(\angle B + \beta)}.$$

Similar reasoning yields

$$\frac{\sin(\angle BAU)}{\sin(\angle UAC)} = \frac{\sin(\gamma)\,\sin(\angle B + \beta)}{\sin(\beta)\,\sin(\angle C + \gamma)} \quad \text{and} \quad \frac{\sin(\angle CBV)}{\sin(\angle VBA)} = \frac{\sin(\alpha)\,\sin(\angle C + \gamma)}{\sin(\gamma)\,\sin(\angle A + \alpha)},$$

and so when we multiply these three ratios of sines to compute the angular Cevian product, everything cancels and the result is 1, as desired. ∎

We mention that the hypothesis in Theorem 4.8 that the three lines all cut across the interior of $\triangle ABC$ is not automatically satisfied, and it can fail. Nevertheless, Theorem 4.8 remains true without this assumption, except that it may turn out that the point of concurrence is "at infinity," which means that lines AU, BV, and CW are parallel, and they are not really concurrent. Also, the theorem continues to hold if we build three inward-pointing triangles on the sides of $\triangle ABC$. Only minor changes in our argument are needed to prove these and other even more general forms of Theorem 4.8, but we shall not pursue these variations any further.

An interesting case of Theorem 4.8 is when $\alpha = \beta = \gamma$ in the notation of Figure 4.9. When that happens, then of course, the three triangles attached to the sides of $\triangle ABC$ are similar isosceles triangles whose bases are the sides of the original triangle. The proof of Theorem 4.8 would be a bit shorter if we assumed this equality of angles because then it would not be necessary to use the law of sines in $\triangle ABW$ to evaluate the ratio AW/BW since the ratio would automatically be 1 in this case. Also, even without the assumption that the three lines AU, BV, and CW cut across the interior of the original triangle, it is not possible for these lines to be parallel in this isosceles situation; they are necessarily concurrent. These lines can be parallel, however, if we allow inward-pointing isosceles triangles.

Let us consider the limiting cases of Theorem 4.8 under the assumption that $\alpha = \beta = \gamma$. If we let these angles approach 0, then points U, V, and W approach the midpoints of sides BC, CA, and AB, respectively. It follows that AU, BV, and CW approach the medians of $\triangle ABC$, and the point of concurrence approaches the centroid of the triangle. The fact that the medians of a triangle are concurrent can thus be viewed as a limiting case of Theorem 4.8. At the other extreme, we can let the equal angles approach 90°. As W recedes, the three lines AW, BW, and CW approach parallelism, with AW and BW approaching perpendicularity to AC. The limit of line CW is thus an altitude of $\triangle ABC$, and the concurrence point approaches the orthocenter of the triangle. The fact that the altitudes of a triangle are concurrent can thus also be viewed as a limiting case of Theorem 4.8.

There is another case of Theorem 4.8 that we knew previously: when all three of α, β, and γ equal 60° and the three attached triangles are actually equilateral. In that case, if each angle of $\triangle ABC$ is less than 120° so that we are in the situation of Figure 4.9, the concurrence point is the Fermat point of the triangle, the unique point where the sum of the distances to the three vertices is a minimum. (See Theorem 2.46.)

As another application of Theorem 4.7, we offer the following.

(4.9) PROBLEM.   In Figure 4.10, the pedal triangle of acute $\triangle ABC$ is $\triangle DEF$, and perpendiculars AU, BV, and CW are dropped from the vertices of the original triangle to the sides of the pedal triangle. Show that lines AU, BV, and CW are concurrent and determine the point of concurrence.

Solution.   By Ceva's theorem for angles, it is enough to check that the angular Cevian product is trivial, and so we want to show that

$$\frac{\sin(\angle ACW)}{\sin(\angle WCB)}\frac{\sin(\angle BAU)}{\sin(\angle UAC)}\frac{\sin(\angle CBV)}{\sin(\angle VBA)} = 1.$$

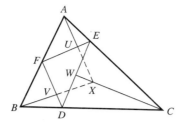

Figure 4.10

Since $\triangle EWC$ is a right triangle, we see that $\sin(\angle ACW) = \cos(\angle WEC)$, and similarly, $\sin(\angle UAC) = \cos(\angle UEA)$. We recall, however, that since $\triangle DEF$ is the pedal triangle of $\triangle ABC$, we have $\angle UEA = \angle WEC$, which is immediate from Corollary 2.43. It follows that $\sin(\angle ACW) = \sin(\angle UAC)$, and similarly, the other factors in the angular Cevian product all cancel, and the lines are concurrent at some point X.

Since $\angle CAX$ and $\angle ACX$ are complementary to the equal angles $\angle AEU$ and $\angle CEW$, we have $\angle CAX = \angle ACX$, and thus $\triangle AXC$ is isosceles and $AX = CX$. Similarly, $BX = CX$, and thus X must be the circumcenter of $\triangle ABC$. ∎

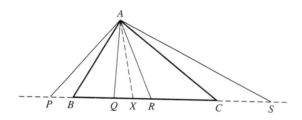

Figure 4.11

Given any Cevian in $\triangle ABC$, there is a natural way to construct a new Cevian from it: Simply reflect it in the bisector of the appropriate angle of the triangle. The resulting line is called the **isogonal** of the original Cevian with respect to the given triangle.

In Figure 4.11, line AX is the bisector of $\angle A$, and lines AQ and AR are the images of each other upon reflection in the bisector AX. In other words, $\angle QAX = \angle RAX$, or equivalently, $\angle QAB = \angle RAC$. Also, AP and AS are reflections of each other in AX, and so $\angle PAX = \angle SAX$ and $\angle PAB = \angle SAC$. In this situation, AQ and AR are isogonals of each other, as are AP and AS. Note also that the isogonal of any interior Cevian is again an interior Cevian, and the isogonal of an exterior Cevian is an exterior Cevian.

The following striking result is an immediate consequence of Theorem 4.7.

(4.10) COROLLARY. *If Cevians AP, BQ, and CR of $\triangle ABC$ are concurrent or parallel, then their isogonal Cevians are also concurrent or parallel.*

Proof.   Observe that the angular Cevian product of the isogonals of three Cevians is
exactly the reciprocal of the angular Cevian product of the original Cevians. Also,
taking isogonals does not change the number of interior Cevians among the three.
If AP, BQ, and CR are concurrent or parallel, then an odd number of them are
interior, and their angular Cevian product is trivial. The same is therefore true about
the isogonals of these three Cevians, and the result follows.   ∎

For example, the isogonals of the medians of $\triangle ABC$ are called the **symmedians** of
the triangle, and they are necessarily concurrent since the medians are concurrent. (The
symmedians cannot be parallel because the medians are interior Cevians, and thus the
symmedians are interior too.) The point of concurrence of the symmedians of a triangle
is called the **Lemoine point**.

Given $\triangle ABC$, choose any point X that is not on one of the lines AB, BC, or
CA. The three lines AX, BX, and CX are Cevians of $\triangle ABC$, and they are obviously
concurrent at the point X. By Corollary 4.9, the isogonals of these Cevians are either
concurrent or parallel, and if they are concurrent at some point Y, then Y is called
the **isogonal conjugate** of X. (In this situation, it should be clear that X is also the
isogonal conjugate of Y.) For example, the centroid and the Lemoine point are isogonal
conjugates of each other. If X is in the interior of the triangle, then the three Cevians
AX, BX, and CX are interior, and thus their isogonal Cevians are interior. Hence they
are necessarily concurrent since they cannot be parallel. It follows that every point in
the interior of the triangle has an isogonal conjugate, and that conjugate is also in the
interior of the triangle. It is easy to see that the incenter of the triangle is the only point
in the interior that is its own isogonal conjugate.

As we have observed, a point X may not have an isogonal conjugate because the
isogonals of the three Cevians AX, BX, and CX may be parallel rather than concurrent.
In fact, this happens when X lies on the circumcircle of $\triangle ABC$.

(4.11) THEOREM.   *Let X be a point other than A, B, or C on the circumcircle of
$\triangle ABC$. Then the isogonal Cevians of AX, BX, and CX are parallel.*

Proof.   It is no loss to assume that X lies between points A and B on the circumcircle,
as shown in Figure 4.12. We have drawn AY and BZ so that $\angle CAY = \angle BAX$ and
$\angle ABX = \angle CBZ$, and for convenience, we have labeled these angles α and β. Thus
Cevian AY is the isogonal of Cevian AX, and Cevian BZ is the isogonal of Cevian

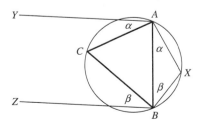

Figure 4.12

BX, and we will show that AY and BZ are parallel. Note that by Corollary 4.10, we know that if AY and BZ are parallel, then the isogonal of Cevian CX is guaranteed to be parallel to these two lines, and so there is no need to consider it any further.

Since the sum of the angles in $\triangle AXB$ is $180°$, we see that $\alpha + \beta = 180° - \angle X = \angle C$, where the second equality holds because quadrilateral $AXBC$ is inscribed in a circle. It follows that

$$\angle YAB + \angle ABZ = \alpha + \angle CAB + \angle ABC + \beta = \angle C + \angle CAB + \angle ABC = 180°$$

Since $\angle YAB$ and $\angle ABZ$ are supplementary, it follows that AY and BZ are parallel, as required.   ∎

In fact, it is not hard to see that it is only for points X on the circumcircle that the three Cevians isogonal to AX, BX, and CX can be parallel. It follows that an isogonal conjugate point is defined for every point X in the plane that is neither on one of the lines AB, BC, or CA nor on the circumcircle. As we see in Figure 4.13, if we exclude these three lines and circle, the remainder of the plane is divided into ten regions that we have labeled a, a', b, b', c, c', w, x, y, and z.

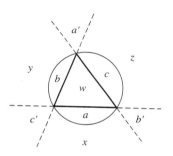

Figure 4.13

A little experimentation shows that the isogonal conjugate of any point in one of the regions w, x, y, or z is back in the same region. (We already knew this, of course, in the case of region w, which is the interior of the triangle.) The points of regions a and a', however, are interchanged by the process of isogonal conjugation, as are the points of regions b and b' and of regions c and c'.

Although much more is known about symmedians and the Lemoine point, and more generally, about isogonal conjugates, we shall give only one further result in this direction.

(4.12) THEOREM. *Let X be any point in the interior of $\triangle ABC$ and let Y be the circumcenter of the triangle whose vertices are the reflections of X in the sides of $\triangle ABC$. Then Y is the isogonal conjugate of X with respect to $\triangle ABC$.*

Proof. It suffices to show that AY is the isogonal Cevian of AX. Similar reasoning would then show that BY and CY are the isogonals of BX and CX, and it would follow that Y is the isogonal conjugate of X. What we must prove, therefore, is that $\angle BAX = \angle CAY$.

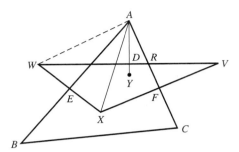

Figure 4.14

Let W and V be the reflections of X in AB and AC, respectively, as shown in Figure 4.14. Then Y is the circumcenter of a triangle having WV as a side, and it follows that the perpendicular bisector of WV passes through Y. Also, AB is the perpendicular bisector of XW, and AC is the perpendicular bisector of XV. So it follows that the point A, where AB and AC meet, is the circumcenter of $\triangle XWV$, and thus A lies on the perpendicular bisector of WV. Since Y also lies on the perpendicular bisector of WV, it follows that AY is this perpendicular bisector, and it meets WV at its midpoint D.

Next, note that $\triangle AWE \cong \triangle AXE$ since AB is the perpendicular bisector of WX. Thus $\angle EAW = \angle EAX$, and we see that $\angle BAX$ is half of $\angle WAX$. We now consider arcs of the circumcircle of $\triangle XWV$, centered at A. Writing R to denote the point where AC meets WV, we compute that

$$\angle BAX = \frac{1}{2}\angle WAX \overset{\circ}{=} \frac{1}{2}\overset{\frown}{WX} \overset{\circ}{=} \angle V = 90° - \angle VRF = 90° - \angle ARD$$
$$= \angle RAD = \angle CAY$$

as desired.   ∎

This section began with a transition from lengths to angles. Starting from Ceva's theorem, which is a concurrence criterion based on lengths, we derived an analogous criterion based on angles. We end with a transition from angles back to lengths. Using the angular Cevian product, we derive another length-based concurrence criterion. This time, however, we consider the diagonals of an inscribed hexagon.

(4.13) THEOREM.   *Let $ABCDEF$ be a hexagon inscribed in a circle. Then the diagonals AD, BE, and CF are concurrent if and only if*

$$\frac{AB}{BC}\frac{CD}{DE}\frac{EF}{FA} = 1.$$

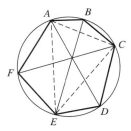

Figure 4.15

Proof.   Draw lines AC, CE, and EA, as shown in Figure 4.15, and note that we can view the three given diagonals of the hexagon as Cevians of $\triangle ACE$. By Theorem 4.7, these diagonals are concurrent if and only if

$$\frac{\sin(\angle AEB)}{\sin(\angle BEC)} \frac{\sin(\angle CAD)}{\sin(\angle DAE)} \frac{\sin(\angle ECF)}{\sin(\angle FCA)} = 1 \, .$$

It suffices, therefore, to show that this angular Cevian product is equal to the hexagonal Cevian product in the statement of the theorem.

By the extended law of sines (Theorem 2.3) in $\triangle ABE$, we have

$$\frac{AB}{\sin(\angle AEB)} = 2R \, ,$$

where R is the radius of the given circle, and hence is the circumradius of $\triangle ABE$. Thus

$$\sin(\angle AEB) = \frac{AB}{2R}$$

and similarly, using the extended law of sines in $\triangle BEC$, we get

$$\sin(\angle BEC) = \frac{BC}{2R} \, .$$

Thus

$$\frac{\sin(\angle AEB)}{\sin(\angle BEC)} = \frac{AB}{BC} \, ,$$

and similarly, the other two ratios of sines that appear in the angular Cevian product are equal to the other two ratios of side lengths of the hexagon in the hexagonal Cevian product. The angular Cevian product is therefore equal to the hexagonal Cevian product, and the proof is complete.   ■

Exercises 4C

4C.1 Given an acute angled $\triangle ABC$, show that the lines joining A, B, and C to the midpoints of the nearer sides of the pedal triangle are concurrent.

4C.2 Show that the only point in the interior of $\triangle ABC$ that is its own isogonal conjugate is the incenter and find all other points in the plane that are their own isogonal conjugates.

4C.3 If $\triangle ABC$ is not a right triangle, show that its circumcenter and orthocenter are isogonal conjugates.

4C.4 If $\angle C = 90°$ in $\triangle ABC$, show that the Lemoine point of the triangle lies on the altitude from vertex C.

4C.5 If the Lemoine point of $\triangle ABC$ lies on the altitude from vertex C, show that either $AC = BC$ or $\angle C = 90°$.

4D   Menelaus' Theorem

Given $\triangle ABC$, let P, Q, and R be arbitrary points on lines BC, CA, and AB, respectively, and assume that none of these points is a vertex of the triangle. If the points P, Q, and R are somehow chosen at random, it is possible, but highly unlikely, that the three Cevians AP, BQ, and CR will be concurrent or parallel. The purpose of Ceva's theorem is to tell us when this miracle occurs: The Cevians will be concurrent or parallel precisely when the Cevian product is trivial and the number of interior Cevians among the three is odd. Note that the condition on the interior Cevians can easily be expressed in terms of the three random points. This oddness condition is exactly equivalent to saying that the number of points in the set $\{P, Q, R\}$ that lie on actual sides of the triangle is odd.

If we choose P, Q, and R at random, as in the previous paragraph, there is another miracle that could happen: The three points might be collinear. Note that this cannot happen if each of P, Q, and R lies on an actual side of the triangle rather than on an extension of the side, and in fact, some experimentation shows that for P, Q, and R to be collinear, it is necessary that either exactly two of them or none of them lie on sides of the triangle. These two possibilities are illustrated, respectively, by the left and right diagrams of Figure 4.16, where we have drawn dashed lines to indicate the extensions of the sides. In other words, the number of members of the set $\{P, Q, R\}$ that lie on sides of the triangle must be even for these points to be collinear. This is, of course, the exact opposite of the situation in Ceva's theorem.

A classical theorem of Menelaus, proved about 100 A.D., tells us when the miracle of collinearity happens. Given that the number of P, Q, and R lying on actual sides

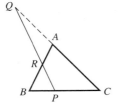
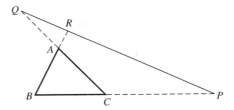

Figure 4.16

of the triangle is even, the condition for collinearity is that the Cevian product is equal to 1. In other words, and rather amazingly, exactly the same numerical condition that is equivalent to the concurrence of the Cevians is also equivalent to the collinearity of the points. The only difference in the conditions for Ceva's and Menelaus' theorems is that for the former, an odd number of members of the set $\{P, Q, R\}$ lie on sides of the triangle, while for the latter, an even number do.

(4.14) THEOREM (Menelaus). *Given* $\triangle ABC$, *let points* P, Q, *and* R *lie on lines* BC, CA, *and* AB, *respectively, and assume that none of these points is a vertex of the triangle. Then* P, Q, *and* R *are collinear if and only if an even number of them lie on segments* BC, CA, *and* AB *and*

$$\frac{AR}{RB}\frac{BP}{PC}\frac{CQ}{QA} = 1.$$

We should mention that in some geometry books, Ceva's and Menelaus' theorems are stated somewhat differently from the way we have presented them here. In those works, the ratios that occur in the Cevian product, such as AR/RB, are sometimes considered to be negative numbers. Specifically, AR/RB is negative if R does not lie between A and B, and it is positive otherwise. With that convention, we see that in the situation of Ceva's theorem, where an odd number of the Cevians are interior, an even number of the ratios AR/RB, BP/PC, and CQ/QA are negative. In the case of Menelaus' theorem, on the other hand, an odd number of these ratios are negative. Using this scheme, the concurrence condition of Ceva is that the Cevian product should equal $+1$, while the collinearity condition of Menelaus is that it should be -1. But we prefer to consider all lengths and ratios of lengths to be positive, and so our statements of Ceva's and Menelaus' theorems refer to oddness and evenness and not to positivity and negativity.

Proof of Theorem 4.14.   First, assume that P, Q, and R are collinear. We can thus assume that we are in one of the two situations depicted in Figure 4.16, and in particular, this forces the number of members of the set $\{P, Q, R\}$ that are on sides of the triangle to be even. We need to show that the Cevian product is trivial.

Draw AP and CR, as shown in Figure 4.17, and work with both diagrams

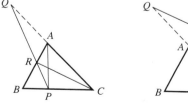

Figure 4.17

simultaneously. Observe that BP and PC can be viewed as the bases of $\triangle BPR$ and $\triangle CPR$, which have equal heights. This yields

$$\frac{BP}{PC} = \frac{K_{BPR}}{K_{CPR}},$$

and similarly, using $\triangle APR$ and $\triangle BPR$, with bases AR and RB, we get

$$\frac{AR}{RB} = \frac{K_{APR}}{K_{BPR}}.$$

We compute the ratio CQ/QA twice, using two pairs of triangles having these segments as bases, namely, $\triangle CQP$ and $\triangle AQP$, and also $\triangle CQR$ and $\triangle AQR$. Thus

$$\frac{CQ}{QA} = \frac{K_{CQP}}{K_{AQP}} = \frac{K_{CQR}}{K_{AQR}},$$

and we can apply the subtraction principle for ratios to get

$$\frac{CQ}{QA} = \frac{K_{CQP} - K_{CQR}}{K_{AQP} - K_{AQR}} = \frac{K_{CPR}}{K_{APR}}.$$

Everything now cancels when we compute $(AR/RB)(BP/PC)(CQ/QA)$, and thus this Cevian product is equal to 1, as desired.

We now assume that the Cevian product is trivial and that an even number of P, Q, and R lie on sides of the triangle. We sketch a proof that these points are collinear, omitting a number of details. A little experimentation with diagrams shows that the only way it can happen that $PQ \parallel AB$, $QR \parallel BC$, and $RP \parallel CA$ is for all three of P, Q, and R to lie on sides of the triangle, and we know that is not the case. We can assume, therefore, that PQ is not parallel to AB, and we let R' be the point where PQ meets AB. Our goal now is to show that R and R' are actually the same point.

The proof proceeds almost exactly as for Ceva's theorem. It is easy to see that R' is neither A nor B and that it lies between A and B if and only if R lies between A and B. The latter assertion holds because even numbers of points in each of the sets $\{P, Q, R\}$ and $\{P, Q, R'\}$ lie on sides of the triangles, and so the numbers for the two sets cannot differ by exactly 1. By hypothesis, the Cevian product for points P, Q, and R is trivial, and also, since P, Q, and R' are collinear, the new Cevian product that results when R is replaced by R' must also be trivial. (We are using the first part of the proof for this, of course.) It follows that $AR/RB = AR'/R'B$, and thus by Lemma 4.3, we conclude that R and R' must be the same point, as desired.   ∎

As an application of Menelaus' theorem, we offer the following.

(4.15) PROBLEM.   In Figure 4.18, we have drawn the tangent lines to the circumcircle of $\triangle ABC$ at the vertices of the triangle. The tangent at A meets line BC at point P, and similarly, the tangents at B and C meet lines CA and AB at points Q and R, respectively. Show that points P, Q, and R are collinear.

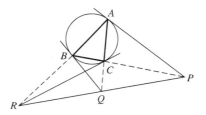

Figure 4.18

Solution.   Since points P, Q, and R lie outside of the circle, none of them lies on a side of the triangle, and thus Menelaus' theorem applies. It suffices, therefore, to compute the three ratios AR/RB, BP/PC, and CQ/QA and show that their product is equal to 1. As we shall see, it is easy to express these three ratios in terms of the lengths a, b, and c of the sides of the original triangle.

We have $\angle BAQ \stackrel{\circ}{=} \frac{1}{2}\,\widehat{BC} \stackrel{\circ}{=} \angle CBQ$, where the second equality follows by Theorem 1.23. Since $\angle BQA = \angle CQB$, we see that $\triangle BAQ \sim \triangle CBQ$ by AA. It follows that

$$\frac{CQ}{BQ} = \frac{BQ}{AQ} = \frac{CB}{BA} = \frac{a}{c},$$

and thus $CQ = (a/c)BQ$ and $AQ = (c/a)BQ$. This yields $CQ/QA = (a/c)/(c/a) = a^2/c^2$. Similarly, $AR/RB = b^2/a^2$ and $BP/PC = c^2/b^2$. It follows that the Cevian product, which is the product of these three quantities, is equal to 1, and the points are collinear, as required.   ∎

The line through P, Q, and R in Problem 4.15 is sometimes called the **Lemoine axis** of the triangle. It has the property that it is perpendicular to the line through the circumcenter and the Lemoine point, although we will not present a proof of that fact here.

Note that one or more of the points P, Q, and R of Problem 4.15 can fail to exist. It may happen, for instance, that the tangent line at A is parallel to BC so that P is undefined. In this case, it is easy to prove that $AB = AC$, and so the triangle is isosceles. If the triangle is not actually equilateral, then Q and R are defined, and it follows that QR is parallel to BC. What is happening, in other words, is that if P does not exist, then the tangent line at A and lines BC and QR are parallel, while if P does exist, these three lines are concurrent at P. Again, as in Ceva's theorem, we see that when three lines are parallel, we have a kind of limiting case of the situation where the three lines are concurrent.

As another application of Menelaus' theorem, we present a theorem of Pappus, who lived in the fourth century. As we will explain, Pappus' theorem is different in flavor from almost everything else in this book.

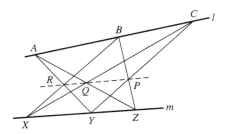

Figure 4.19

(4.16) THEOREM (Pappus).   *Suppose that points A, B, and C lie on some line l and that points X, Y, and Z lie on line m, where the six points are distinct and the two lines are also distinct. Assume that lines BZ and CY meet at P, lines AZ and CX meet at Q, and lines AY and BX meet at R. Then points P, Q, and R are collinear.*

Note that Figure 4.19 illustrates just one of the many possible configurations for Pappus' theorem. We deliberately drew lines l and m to be nearly parallel, and we placed the points A, B, and C on l and X, Y, and Z on m in the orders shown because that arrangement forces the three allegedly collinear points to remain nearby, and this keeps the entire diagram from becoming unmanageably large. We stress, however, that the two lines and the six points need not be arranged in this way; they are completely arbitrary except for the conditions stated in the theorem. Note, however, that if one of the six points happens to be the intersection point of the two lines, then points P, Q, and R will not be distinct. The theorem is trivially true in that case because two points are automatically collinear.

Pappus' theorem is true somewhat more generally than we have stated it. Suppose, for example, that lines BZ and CY happen to be parallel so that point P does not exist. (Of course, this cannot happen if the points are arranged as in Figure 4.18, but BZ and CY certainly can be parallel in other configurations.) We can then work in the extended projective plane and take P to be the imaginary ideal point at infinity, where the parallel lines BZ and CY meet. To say that P is collinear with Q and R in this situation means that line QR contains the ideal point P, and this tells us that QR is parallel to BZ and CY. We will not prove the versions of Pappus' theorem involving ideal points and parallel lines, but it is a fact that with suitable interpretation, all such generalizations are true.

Pappus' theorem has a different flavor from any of the other results we have studied because it is entirely nonmetric. To understand this, suppose that in some particular situation, we want to check that the hypotheses of Pappus' theorem hold, or we want to confirm that its conclusion is valid. All that we need to do is check that certain points lie on certain lines. What is needed for Pappus' theorem are only the notions of point, line, and incidence. (We say that a point and line are **incident** if the point lies on the line, or equivalently, the line goes through the point.) The lengths of line segments and the sizes of angles are completely irrelevant for Pappus' theorem. There is absolutely nothing to be measured, and so we refer to Pappus' theorem and other results of this type as belonging to the area of **nonmetric geometry**. Note that no result involving circles could be called nonmetric because a circle is defined as the locus of points of some fixed distance from a given point, and distance is, of course, a metric concept.

Since Pappus' theorem is about points, lines, incidence, and absolutely nothing else, we can also say that this result belongs to **projective geometry**. To explain this term, we imagine that Figure 4.19 or some other diagram illustrating Pappus' theorem is drawn with opaque ink on a sheet of glass and that a point source of light causes the figure to cast a shadow onto a planar screen. Since this projection from a point carries points to points and lines to lines, and it preserves incidence, we see that the shadow of a diagram for Pappus' theorem is again a diagram for Pappus' theorem. In a very rough sense, projective geometry is that part of ordinary (Euclidean) geometry where the shadows of diagrams illustrate the relevant information in the original diagrams. Thus Pappus' theorem belongs to projective geometry, but the pons asinorum, for example, does not because the shadow (projection) of an isosceles triangle need not be isosceles. Also, Ceva's and Menelaus' theorems should probably not be considered as belonging to projective geometry even though concurrence of Cevians and collinearity of points are preserved by projections. The nonprojective aspect of these results is that they concern ratios such as BP/PC, which are not preserved. Note, however, that cross ratios are preserved; this is the essential content of Theorem 3.15.

Although we have just asserted that Pappus' theorem is unlike anything that we have seen before, there is, in fact, a closely related result that appeared in Exercise 3C.7. In that theorem of Pascal, we also have six points that are joined by three pairs of lines intersecting in three points, and there too the conclusion is that the three intersection points are necessarily collinear. In Pascal's theorem, however, the six points lie on a circle, whereas in Pappus' theorem, they lie on two lines. Note that, as we stated it, Pascal's theorem is neither nonmetric nor projective because it involves a circle. Projections of circles are ellipses and other conic sections, however, and so if we restate Pascal's theorem with an arbitrary conic section instead of a circle, we get a version of the result that really is a theorem of projective geometry. Also, it should be noted that the truth of this more general theorem follows from the special case for circles by projecting the appropriate diagram.

The proof of Pascal's theorem that was suggested in the hint for Exercise 3C.7 used cross ratios, and in fact, an entirely analogous proof is available for Pappus' theorem. We prefer, however, to deduce the Pappus result from Menelaus' theorem. The proof that follows is not quite complete, however, because it assumes that certain lines are not parallel. One could prove the full result from the case that we consider by using limit arguments, but we will omit this refinement. A better argument that avoids the consideration of such special cases is also available, using techniques of linear algebra, but we shall not pursue that further here.

Proof of Theorem 4.16. Define point L to be the intersection of XC with AY, as shown in Figure 4.20, and similarly, let M be the intersection of AY with BZ and let N be the intersection of BZ with XC. Of course, we are assuming that none of the pairs of lines defining points L, M, and N is parallel so that these three points are actually defined.

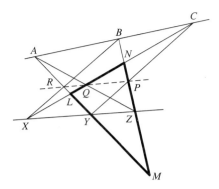

Figure 4.20

Note that points P, Q, and R lie on lines MN, NL, and LM, respectively. We propose to prove that these three points are collinear by applying Menelaus' theorem to $\triangle LMN$, drawn in Figure 4.20 with heavy ink. We shall compute the Cevian product $(LR/RM)(MP/PN)(NQ/QL)$ and show that it is equal to 1. To complete the proof, we should also check, of course, that the number of members of the set $\{P, Q, R\}$ that lie on actual sides of $\triangle LMN$ is even. This is certainly true in Figure 4.20, where the number is 2, but we shall omit the verification of this in other configurations of the points, and we proceed directly to the computation of the Cevian product.

Observe that points A, B, and C are collinear and lie on lines LM, MN, and NL, respectively, and so we deduce by Menelaus' theorem that

$$\frac{LA}{AM}\frac{MB}{BN}\frac{NC}{CL} = 1,$$

and similarly, from the fact that X, Y, and Z are collinear, we get

$$\frac{LY}{YM}\frac{MZ}{ZN}\frac{NX}{XL} = 1.$$

We have several more triples of collinear points, and so we get additional Cevian products equal to 1. For example, R, B, and X are collinear, and thus

$$\frac{LR}{RM}\frac{MB}{BN}\frac{NX}{XL} = 1.$$

Similarly, from the collinearity of A, Q, and Z and of C, P, and Y, we get, respectively,

$$\frac{LA}{AM}\frac{MZ}{ZN}\frac{NQ}{QL} = 1 \quad \text{and} \quad \frac{LY}{YM}\frac{MP}{PN}\frac{NC}{CL} = 1.$$

Multiplying the last three equations and the first two, we get

$$\frac{LR}{RM}\frac{MB}{BN}\frac{NX}{XL}\frac{LA}{AM}\frac{MZ}{ZN}\frac{NQ}{QL}\frac{LY}{YM}\frac{MP}{PN}\frac{NC}{CL} = 1 = \frac{LA}{AM}\frac{MB}{BN}\frac{NC}{CL}\frac{LY}{YM}\frac{MZ}{ZN}\frac{NX}{XL}.$$

Now the six fractions on the right cancel with six of the nine fractions on the left, and what results is the equation

$$\frac{LR}{RM}\frac{MP}{PN}\frac{NQ}{QL} = 1,$$

as we wanted.   ∎

Exercises 4D

4D.1 We say that points X and Y are **symmetric points** with respect to a point M if M is the midpoint of segment XY. Given $\triangle ABC$, suppose that l is a line not parallel to any of its sides, and let P, Q, and R be the points of intersection of l with lines BC, CA, and AB, respectively. Suppose that points P and X are symmetric with respect to the midpoint of segment BC. Similarly, assume that Q and Y are symmetric with respect to the midpoint of CA and that R and Z are symmetric with respect to the midpoint of AB. Show that X, Y, and Z are collinear. In the case that l passes through a vertex of the triangle, which line passes through X, Y, and Z?

4D.2 Let AP and BQ be the bisectors of $\angle A$ and $\angle B$ in $\triangle ABC$ and suppose that CR is perpendicular to the bisector of $\angle C$, where R lies on an extension of side AB. Show that points P, Q, and R are collinear.

HINT: Compute the relevant Cevian product using angles, as in Theorem 4.7.

4D.3 Three concurrent Cevians AP, BQ, and CR are drawn in $\triangle ABC$, as shown in Figure 4.21, and RQ is extended to meet the extension of BC at S. Apply both Ceva's and Menelaus' theorem in $\triangle ABC$ to prove that $xz = y(x+y+z)$, where we have written $BP = x$, $PC = y$, and $CS = z$, as indicated.

NOTE: Exercise 4D.3 provides an alternative solution, independent of cross ratios, to Problem 3.16.

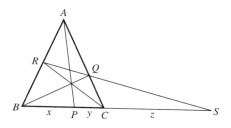

Figure 4.21

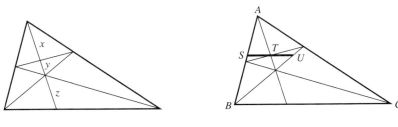

Figure 4.22   Figure 4.23

4D.4 Figure 4.22 shows three concurrent interior Cevians in a triangle. One of these is divided into three parts by the point of concurrence and by the line joining the ends of the other two Cevians. If the three pieces have lengths x, y, and z, as shown, prove that $xz = y(x + y + z)$.

HINT: As in Exercise 4D.3, this can be done by applying both Ceva's and Menelaus' theorems to an appropriate triangle. It is not necessary to draw any extra lines or to extend any of the line segments in the diagram.

4D.5 As shown in Figure 4.23, three concurrent interior Cevians are drawn in $\triangle ABC$, and the intersection of the Cevian from A with the line joining the ends of the other two Cevians is denoted T. Line segment SU is drawn through T, parallel to BC, as shown, where S lies on side AB and U lies on the Cevian from B. Prove that $ST = TU$.

4D.6 Figure 4.24 shows three circles of different sizes, none of which is inside any of the others. For each pair of these circles, the two common exterior tangents are drawn, and these three pairs of tangents are extended to meet at points P, Q, and R. Prove that these three points are collinear.

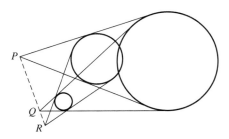

Figure 4.24

4D.7 In Figure 4.25, we see that $\triangle ABC$ and $\triangle XYZ$ are in **perspective** from point S, which means that lines AX, BY, and CZ all go through point S. The corresponding sides of these triangles, when extended, meet at points P, Q, and R, as shown. Prove that these three points are collinear, as indicated by the dashed line.

HINT: Apply Menelaus' theorem a total of four times. In $\triangle SBC$, the Cevian product for the collinear points P, Z, and Y is trivial. Two more trivial Cevian

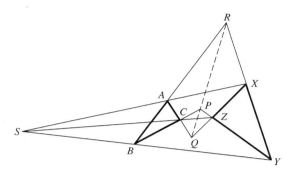

Figure 4.25

products can be obtained from $\triangle SCA$ and SAB. Multiply the three Cevian products thus obtained.

NOTE: The result of this problem is a theorem of Girard Desargues (1593–1662). Observe that like Pappus' theorem, Desargues' theorem is entirely nonmetric and belongs to the area of projective geometry.

CHAPTER FIVE

Vector Methods of Proof

5A Vectors

In this chapter, we show how vectors can be used to prove some striking geometry theorems that are difficult to prove using more conventional techniques. We begin with a brief review of the definition of vectors and of some of their basic properties.

First, we introduce our notation. To distinguish vectors from other types of objects, such as numbers or points, they are generally written like this: $\vec{v}$ or $\vec{A}$ or $\vec{AB}$, with a little arrow over the symbol or symbols. But what is a vector? A plane vector $\vec{v}$ is simply an ordered pair of real numbers, which are called its **coordinates**. We can thus write $\vec{v} = (a, b)$, where the coordinates a and b are just numbers. Although in coordinate geometry, such pairs of real numbers are used to denote points in the plane, we prefer not to think of vectors as points; vectors are just pairs of numbers. If one is working in three dimensional space and is not confined to a plane, it is convenient to use space vectors, which are defined to be ordered triples (a, b, c) of real numbers. In fact, it is often useful to consider n-dimensional vectors that look like $(a_1, a_2, a_3, \ldots a_n)$, where there are n coordinates instead of just two or three and n can be any positive integer. Our purpose in introducing vectors is to use them to prove facts in plane geometry, however, and so we will limit ourselves henceforth to vectors having just two coordinates.

Vectors can be added or subtracted by adding or subtracting the corresponding coordinates. Thus if $\vec{v} = (a, b)$ and $\vec{w} = (c, d)$, we have $\vec{v} + \vec{w} = (a + c, b + d)$ and $\vec{v} - \vec{w} = (a - c, b - d)$. Also, we can multiply vectors by scalars simply by multiplying each coordinate by that scalar. (A **scalar** is nothing but an ordinary real number.) If z is scalar and $\vec{v} = (a, b)$ is a vector, therefore, we write $z\vec{v}$ to denote the vector (za, zb).

Many of the usual rules of arithmetic also hold for vectors. For example, the commutative and associative laws are valid for vector addition, and two distributive laws hold for addition and scalar multiplication. (The two distributive laws are these:

$$(y + z)\vec{v} = y\vec{v} + z\vec{v} \quad \text{and}$$
$$y(\vec{v} + \vec{w}) = y\vec{v} + y\vec{w},$$

where y and z are scalars and $\vec{v}$ and $\vec{w}$ are vectors.) Also, the vector $\vec{0} = (0, 0)$, which is called the **zero vector**, behaves very much like the number 0 in ordinary arithmetic: If $\vec{v}$ is any vector and z is any scalar, then $\vec{v} + \vec{0} = \vec{v}$ and $z\vec{0} = \vec{0}$.

To relate vectors to geometry, we represent each vector as an arrow in the plane. To be specific, let us assume that our plane comes equipped with a coordinate system so that each point P can be described as an ordered pair (x, y). Of course, (x, y) looks just like a vector, but we refuse to think of it as a vector; it is just a way of naming the point P.

Suppose now that we are given a vector $\vec{v} = (a, b)$. Let P be any point in the plane and suppose that its coordinates are (x, y). If we let Q be the point whose coordinates are $(x + a, y + b)$, then we can think of the vector $\vec{v}$ as instructions about how to get from point P to point Q: Go a units right and b units up. Of course, if a is negative, we actually move left, and if b is negative, we move down. If we draw an arrow from P to Q with tail at P and head at Q, then we think of this arrow as being a picture of the vector $\vec{v}$, and we write $\overrightarrow{PQ} = \vec{v}$. Often, we think of the arrow from P to Q as actually being the vector $\vec{v}$, but this can be dangerous because it is essential to remember that the point P was chosen arbitrarily; it was not in any sense determined by the vector $\vec{v}$. But, of course, once P is chosen, then Q is unambiguously determined. The vector $\vec{v}$ is represented by infinitely many different arrows in the plane: one for each choice of the tail point P. We can think of any one of these arrows as being a picture or representation of $\vec{v}$, and we shall see that all of the arrows that represent $\vec{v}$ are parallel and have equal lengths. Each of them can be obtained from any of the others by a translation, which is a motion in the plane without rotation.

We have a slight problem if $\vec{v}$ is the zero vector $\vec{0} = (0, 0)$, since in that case, the points P and Q are identical, and we cannot draw an actual arrow from P to Q. Nevertheless, $\vec{0}$ does give instructions for how to get from P to P, and we can at least imagine a corresponding arrow with zero length and no particular direction.

Given two points P and Q and an arrow with tail at P and head at Q, we can reconstruct the vector $\vec{v} = \overrightarrow{PQ}$ by subtracting the corresponding coordinates of $P = (x_1, y_1)$ and $Q = (x_2, y_2)$. Thus $\overrightarrow{PQ} = (x_2 - x_1, y_2 - y_1)$, and we see that every arrow we can draw represents some vector. Note that we need the arrow from P to Q and not just the line segment PQ, because we need to know which point is the head and which is the tail so that we can subtract the tail coordinates from the head coordinates, and not vice versa. In fact, we see that $\overrightarrow{QP} = -\overrightarrow{PQ}$.

What is the geometric meaning of vector addition? Given vectors $\vec{v}$ and $\vec{w}$, we represent $\vec{v}$ as an arrow from P to Q, where P is arbitrary. Although we can represent $\vec{w}$ as an arrow with any starting point (tail) that we like, we choose to draw $\vec{w}$ starting from Q, and we write $\vec{w} = \overrightarrow{QR}$. We have thus placed the arrows representing v and w with the head of $\vec{v}$ at the tail of $\vec{w}$. It is easy to see that the arrow from P to R represents $\vec{v} + \vec{w}$. In other words, we have the vector equation $\overrightarrow{PQ} + \overrightarrow{QR} = \overrightarrow{PR}$. We can also think of this in the following way: The instructions for going from P to R are first to go from P to Q and then to go from Q to R.

We have already remarked that any given vector can be represented by infinitely many different arrows. Given the four points P, Q, R, and S, suppose it happens that $\overrightarrow{PQ} = \overrightarrow{RS}$. We mentioned previously that in this case, line segments PQ and RS must be equal and parallel. To see why this is true, consider Figure 5.1, where we have drawn right triangles $\triangle PQX$ and $\triangle RSY$ with horizontal and vertical arms and with our given equal vectors as hypotenuses. (We really should say, of course, that the arrows

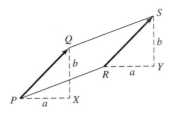

Figure 5.1

representing the vector are the hypotenuses, but it is convenient to talk about the arrows as though they actually were the vectors.)

If we write $\overrightarrow{PQ} = (a, b) = \overrightarrow{RS}$, we see that $PX = a = RY$ and $XQ = b = YS$, and thus the two right triangles are congruent by SAS. (For simplicity, we are working in the case where the coordinates a and b are positive, but this is not really essential.) It follows that the lengths PQ and RS are equal. In fact, by the Pythagorean theorem, we see that the lengths of $\overrightarrow{PR}$ and $\overrightarrow{QS}$ are both equal to $\sqrt{a^2 + b^2}$. This shows that two arrows representing the same vector must have equal lengths. But more is true. We have

$$\overrightarrow{PQ} + \overrightarrow{QS} = \overrightarrow{PS} = \overrightarrow{PR} + \overrightarrow{RS},$$

and if we subtract the equal vectors $\overrightarrow{PQ} = \overrightarrow{RS}$ from both sides, we deduce that $\overrightarrow{QS} = \overrightarrow{PR}$. It follows from what we just proved that the corresponding arrows have equal lengths, and so we can write $QS = PR$. We conclude that quadrilateral $PQSR$ is a parallelogram, and hence $PQ \| RS$. This shows that all arrows representing the same vector are equal and parallel, as we claimed. Conversely, it is not hard to see that two arrows that are equal, parallel, and point in the same rather than in opposite directions correspond to equal vectors.

Finally, we mention that the geometric significance of multiplication of a vector $\vec{v}$ by a positive scalar z is that an arrow representing $z\vec{v}$ points in the same direction as an arrow representing $\vec{v}$, but it is shorter than, equal to, or longer than the original arrow according to whether z is less than, equal to, or greater than 1. More precisely, the length of $z\vec{v}$ is exactly z times the length of $\vec{v}$. If the scalar z is negative, the direction of the vector is reversed, but otherwise, we get the same shrinking or stretching effect as with a positive scalar. For example, an arrow representing $-3\vec{v}$ has three times the length of an arrow representing $\vec{v}$, but it points in the opposite direction.

To see some further examples, suppose that P, Q, R, and S are four points lying in that order along a line and assume that they are equally spaced so that $PQ = QR = RS$. Then $\overrightarrow{PQ} = \overrightarrow{QR} = \overrightarrow{RS}$ and $\overrightarrow{PR} = \overrightarrow{QS}$. Some of the other equations that we can write in this situation are $-\overrightarrow{SP} = \overrightarrow{PS} = 3\overrightarrow{PQ}$ and $\overrightarrow{PR} = -\frac{2}{3}\overrightarrow{SP}$.

5B   Vectors and Geometry

For convenience in applying vector techniques to geometry, we introduce a notational shortcut. We suppose that some point O, which we call the **origin**, has been selected in the plane, and we hold this point fixed. As we will see, it is not usually necessary to know the actual position of this point, but sometimes it is possible to simplify a proof by choosing the origin O in some particularly clever way. The notational shortcut to which we referred is that a vector of the form $\overrightarrow{OA}$, with tail at point O, will simply be written as $\vec{A}$. In other words, whenever a vector is named by a single point rather than by a pair of points, we assume that the tail of the corresponding arrow is at the origin and that the head is at the named point.

Given two points A and B in the plane, we know that $\overrightarrow{OA} + \overrightarrow{AB} = \overrightarrow{OB}$, and using the notational shortcut just described, we can rewrite this as $\vec{A} + \overrightarrow{AB} = \vec{B}$. From this, we get $\overrightarrow{AB} = \vec{B} - \vec{A}$, and hence any vector named by two points can be described as a difference of two vectors, each of which is named by a single point. Notice that the correct way to do this is always head minus tail. The vector from P to Q, for example, is $\vec{Q} - \vec{P}$. We mention that one way to prove that two points P and Q are actually identical is to show that $\overrightarrow{PQ} = \vec{0}$. But $\overrightarrow{PQ} = \vec{Q} - \vec{P}$, and this is the zero vector precisely when $\vec{Q} = \vec{P}$. In other words, to show that P and Q are the same point, it suffices to show that the vectors $\vec{P}$ and $\vec{Q}$ corresponding to these points are equal.

(5.1) PROBLEM. If point M is the midpoint of line segment AB, show how to express the vector $\vec{M}$ in terms of $\vec{A}$ and $\vec{B}$.

Solution. To get to M from A, we need to travel exactly half of the way from A to B. This can be expressed in vector language by writing $\overrightarrow{AM} = \frac{1}{2}\overrightarrow{AB}$, and using our notational shortcut, we can rewrite this as $\vec{M} - \vec{A} = \frac{1}{2}(\vec{B} - \vec{A})$. Thus $\vec{M} = \vec{A} + \frac{1}{2}(\vec{B} - \vec{A})$, and a bit of algebraic simplification yields $\vec{M} = \frac{1}{2}(\vec{A} + \vec{B})$. ∎

The result of Problem 5.1 is useful and easy to remember. It says that the vector $\vec{M}$ corresponding to the midpoint M of line segment AB is exactly the average of the vectors $\vec{A}$ and $\vec{B}$, corresponding to the endpoints of the segment. And note that we can make this statement without knowing which point was selected to be the origin.

As our first example of a geometry proof using vectors, we give another argument that shows that the three medians of a triangle are concurrent.

(5.2) THEOREM. *The medians of $\triangle ABC$ are concurrent at a point G that lies two thirds of the way along each median (moving from a vertex to the midpoint of the opposite side). Furthermore, $\vec{G} = \frac{1}{3}(\vec{A} + \vec{B} + \vec{C})$.*

Proof. We compute the vector $\vec{G}$ corresponding to the point G that lies two thirds of the way along median AM, where M is the midpoint of BC. By Problem 5.1, we know that $\vec{M} = \frac{1}{2}(\vec{B} + \vec{C})$, and we have

$$\vec{G} - \vec{A} = \overrightarrow{AG} = \frac{2}{3}\overrightarrow{AM} = \frac{2}{3}\left(\vec{M} - \vec{A}\right) = \frac{2}{3}\left(\frac{1}{2}(\vec{B} + \vec{C}) - \vec{A}\right).$$

Some easy algebra now yields that $\vec{G} = \frac{1}{3}(\vec{A} + \vec{B} + \vec{C})$. In other words, the vector corresponding to the point two thirds of the way along median AM is the average of the three vectors corresponding to the vertices of the triangle. Similar reasoning shows that the vector corresponding to the point two thirds of the way along each of the other two medians must also be the average of the three vectors corresponding to the vertices. The vectors corresponding to the points two thirds of the way along the three medians are therefore equal, and it follows that these three points are identical. This completes the proof.   ∎

One disadvantage of this method of proof that the medians of a triangle are concurrent is that in order to use it, we had to know in advance that the point of concurrence lies two thirds of the way along each median. If we hadn't already known or guessed this fact, we could not have found this proof. On the other hand, once we know what we are trying to prove, the method works purely mechanically; we do not need to be clever. We simply compute the vectors corresponding to the three points that are two thirds of the way along the three medians, and these three vectors turn out to be equal. If the theorem is true, then in some sense, this method of proof *must* work.

Here is another easy example.

(5.3) PROBLEM.   Let $ABCD$ be any quadrilateral and let W, X, Y, and Z be the midpoints of AB, BC, CD, and DA, as shown in Figure 5.2. Give a vector proof that $WXYZ$ is a parallelogram.

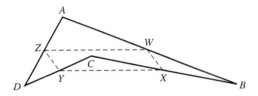

Figure 5.2

Recall that this is a fact we have seen before. Our previous proof required one little trick: Draw AC. Then WX is a line joining the midpoints of two sides of $\triangle ABC$, and thus $WX \| AC$ and $WX = \frac{1}{2}AC$. Similarly, $ZY \| AC$ and $ZY = \frac{1}{2}AC$, and hence WX and ZY are equal and parallel, and the result follows.

Solution to Problem 5.3.   We want to show that $\overrightarrow{WX} = \overrightarrow{ZY}$ since that will imply that WX is both parallel and equal to ZY, and the result will follow. The given data are the arbitrary four points A, B, C, and D, and so we will work to express $\overrightarrow{WX}$ and $\overrightarrow{ZY}$ in terms of $\vec{A}$, $\vec{B}$, $\vec{C}$, and $\vec{D}$. We expect that we will get the same expression for both $\overrightarrow{WX}$ and $\overrightarrow{ZY}$, and if we do, that will complete the proof.
We have $\vec{W} = \frac{1}{2}(\vec{A} + \vec{B})$ and $\vec{X} = \frac{1}{2}(\vec{B} + \vec{C})$. Thus

$$\overrightarrow{WX} = \vec{X} - \vec{W} = \frac{1}{2}(\vec{B} + \vec{C}) - \frac{1}{2}(\vec{A} + \vec{B}) = \frac{1}{2}(\vec{C} - \vec{A}) \ .$$

Similarly,

$$\overrightarrow{ZY} = \vec{Y} - \vec{Z} = \frac{1}{2}(\vec{C} + \vec{D}) - \frac{1}{2}(\vec{D} + \vec{A}) = \frac{1}{2}(\vec{C} - \vec{A}) \ ,$$

and since this is the same expression that we obtained for $\overrightarrow{WX}$, the proof is complete.   ■

We give one more example of a problem that can be solved by this method.

(5.4) PROBLEM.   Given $\triangle ABC$, we construct $\triangle RST$ by taking points R, S, and T on the sides of the original triangle, as follows. Point R lies one third of the way from A to B along AB, point S lies one third of the way from B to C along BC, and point T lies one third of the way from C to A along CA. Now repeat this process starting with $\triangle RST$ and obtain $\triangle XYZ$, as shown in Figure 5.3. Show that $\triangle XYZ \sim \triangle CAB$ and show that the corresponding sides of these two triangles are parallel.

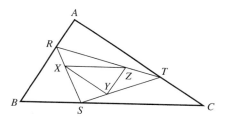

Figure 5.3

To solve Problem 5.4 and other related problems, it is convenient to have a general method for determining the vector corresponding to the point obtained by moving a specified fraction γ of the way along a given line segment AB.

(5.5) LEMMA.   *Let γ be a real number with $0 < \gamma < 1$ and suppose that X is the point lying γ of the way from A to B along segment AB. Then $\vec{X} = (1 - \gamma)\vec{A} + \gamma\vec{B}$.*

For example, suppose that $\gamma = 1/2$. Then X is the point that lies half of the way from A to B, and so X is the midpoint of segment AB. In this case, the lemma asserts that $\vec{X} = \frac{1}{2}\vec{A} + \frac{1}{2}\vec{B}$, and not surprisingly, this agrees with our earlier formula for midpoints. Also note that as γ approaches 0, point X approaches point A, and so $\vec{X}$ should approach $\vec{A}$, and this is consistent with the formula given in the lemma since $1 - \gamma$ approaches 1 as γ approaches 0. Similarly, as γ approaches 1, we see that X approaches B, and this is also consistent with the lemma.

Proof of Lemma 5.5.   We see that $\overrightarrow{AX} = \gamma\overrightarrow{AB}$, and thus $\vec{X} - \vec{A} = \gamma(\vec{B} - \vec{A})$. We compute that

$$\vec{X} = \vec{A} + \gamma(\vec{B} - \vec{A}) = (1 - \gamma)\vec{A} + \gamma\vec{B},$$

as required.   ■

Solution to Problem 5.4.   The given data are the points A, B, and C, and so our strategy is to express the vectors along the sides of $\triangle XYZ$ in terms of $\vec{A}$, $\vec{B}$, and $\vec{C}$. First, since R is one third of the way from A to B, we see by Lemma 5.5 that $\vec{R} = \frac{2}{3}\vec{A} + \frac{1}{3}\vec{B}$. Similarly, $\vec{S} = \frac{2}{3}\vec{B} + \frac{1}{3}\vec{C}$. Since X lies one third of the way from R to S, Lemma 5.5 yields

$$\vec{X} = \frac{2}{3}\vec{R} + \frac{1}{3}\vec{S} = \frac{2}{3}\left(\frac{2}{3}\vec{A} + \frac{1}{3}\vec{B}\right) + \frac{1}{3}\left(\frac{2}{3}\vec{B} + \frac{1}{3}\vec{C}\right).$$

A little elementary algebra now yields $\vec{X} = \frac{4}{9}\vec{A} + \frac{4}{9}\vec{B} + \frac{1}{9}\vec{C}$. Now that we have a formula for $\vec{X}$, we can see what the formula for $\vec{Y}$ must be without doing any real work, which is the preferred method. To get the formula for $\vec{Y}$, simply move around the triangle and replace A by B, B by C, and C by A. This gives $\vec{Y} = \frac{4}{9}\vec{B} + \frac{4}{9}\vec{C} + \frac{1}{9}\vec{A}$, and we have

$$\overrightarrow{XY} = \vec{Y} - \vec{X} = \left(\frac{4}{9}\vec{B} + \frac{4}{9}\vec{C} + \frac{1}{9}\vec{A}\right) - \left(\frac{4}{9}\vec{A} + \frac{4}{9}\vec{B} + \frac{1}{9}\vec{C}\right)$$

$$= \frac{1}{3}\vec{C} - \frac{1}{3}\vec{A} = \frac{1}{3}\overrightarrow{AC}.$$

Since the vector $\overrightarrow{XY}$ is one third of the vector $\overrightarrow{AC}$, we know that the corresponding arrows are parallel and that the former has one third the length of the latter. Thus $XY \| CA$ and $XY = \frac{1}{3}CA$. Similarly, each side of $\triangle XYZ$ is parallel to the corresponding side of $\triangle CAB$, and each side of $\triangle XYZ$ has length equal to one third of the length of the corresponding side of $\triangle CAB$. Thus $\triangle XYZ \sim \triangle CAB$ by SSS, and the proof is complete. ∎

We mention that if we have any two triangles such that the three sides of one are parallel to the three sides of the other, then the triangles are automatically similar. This follows fairly easily by AA.

Exercises 5B

5B.1 Give a vector proof of the fact that if X and Y are points on sides AB and AC of $\triangle ABC$ and $AX/AB = AY/AC$, then $XY \| BC$.

5B.2 In the situation of Problem 5.4, we saw that the sides of $\triangle XYZ$ were each one third of the sides of the similar $\triangle CAB$, and it follows that the area of $\triangle XYZ$ is one ninth of the area of the original $\triangle ABC$. Recall that we obtained $\triangle XYZ$ by applying a certain procedure twice to the starting triangle. The first application yielded $\triangle RST$, and then when the procedure was applied to $\triangle RST$, the result was $\triangle XYZ$. This suggests, but does not prove, that the procedure always yields a triangle with one third of the area of the triangle to which it is applied. Prove that this is true.

5B.3 Fix a number γ with $0 < \gamma < 1$. Starting with $\triangle ABC$, construct $\triangle RST$ by taking R to be γ of the way from A to B on AB, and similarly, take S and T to be γ of the way along sides BC and CA. (If $\gamma = 1/3$, therefore, this is exactly the process of Problem 5.4.) Next, apply a "backward version" of this process to $\triangle RST$ to obtain $\triangle XYZ$. Specifically, take X to be γ of the way from S to R on RS, and similarly, take Y and Z to be γ of the way from T to S and γ of the way from R to T, respectively. (Note that even if $\gamma = 1/3$, this does not yield the same $\triangle XYZ$ as we had in Problem 5.4.) Show that $\triangle XYZ$ is similar to the original triangle with an appropriate ordering of the vertices and that corresponding sides are parallel.

5C Dot Products

Most readers of this book are probably familiar with the dot product $\vec{v} \cdot \vec{w}$ of two vectors $\vec{v}$ and $\vec{w}$. If we write $\vec{v}$ for the ordered pair (a, b) and $\vec{w}$ for the ordered pair (c, d), then the dot product $\vec{v} \cdot \vec{w}$ is defined to be the scalar $ac + bd$. Dot products, which are sometimes called scalar products, can also be defined in three or more dimensions, and the rule is the same in all cases: Multiply the corresponding coordinates and then add the results. (Thus in three dimensions, if $\vec{v} = (a_1, a_2, a_3)$ and $\vec{w} = (b_1, b_2, b_3)$, we have $\vec{v} \cdot \vec{w} = a_1 b_1 + a_2 b_2 + a_3 b_3$.) It is easy to check that the commutative and distributive laws hold for dot products. In other words, if $\vec{u}$, $\vec{v}$, and $\vec{w}$ are any three vectors, we have the following: $\vec{u} \cdot \vec{v} = \vec{v} \cdot \vec{u}$ and also $\vec{u} \cdot (\vec{v} + \vec{w}) = \vec{u} \cdot \vec{v} + \vec{u} \cdot \vec{w}$. Note that in the last equation, the plus sign on the left represents vector addition, but the plus sign on the right represents ordinary scalar addition.

Returning to vectors in the plane, we investigate the geometric significance of the dot product. First, we consider the dot product of a vector with itself. If $\vec{v} = (a, b)$, we see that $\vec{v} \cdot \vec{v} = a^2 + b^2$, and we should recognize this as the square of the length of an arrow representing $\vec{v}$. (Of course, we are appealing to the Pythagorean theorem here.) It is customary to use the absolute value notation to represent the length of a vector, and thus we can write $\vec{v} \cdot \vec{v} = |\vec{v}|^2$. Note that if P and Q are points, and if as usual, we write PQ to denote the length of the line segment they determine, we can write $|\overrightarrow{PQ}| = PQ$.

Now consider $\triangle ABC$ and, as usual, let a, b, and c denote the lengths of sides BC, AC, and AB, respectively. Recall that the law of cosines tells us that $a^2 = b^2 + c^2 - 2bc \cos(A)$. Write $\vec{v} = \overrightarrow{AC}$ and $\vec{w} = \overrightarrow{AB}$ so that we have $\vec{v} \cdot \vec{v} = |\vec{v}|^2 = (AC)^2 = b^2$, and similarly, $\vec{w} \cdot \vec{w} = c^2$. Since

$$\vec{v} = \overrightarrow{AC} = \overrightarrow{AB} + \overrightarrow{BC} = \vec{w} + \overrightarrow{BC} ,$$

we see that $\overrightarrow{BC} = \vec{v} - \vec{w}$, and thus

$$(\vec{v} - \vec{w}) \cdot (\vec{v} - \vec{w}) = \left| \overrightarrow{BC} \right|^2 = (BC)^2 = a^2 .$$

If we compute the left side of the previous equation using the commutative and distributive laws for dot products, we get

$$(\vec{v} - \vec{w}) \cdot (\vec{v} - \vec{w}) = (\vec{v} - \vec{w}) \cdot \vec{v} - (\vec{v} - \vec{w}) \cdot \vec{w}$$
$$= \vec{v} \cdot \vec{v} - \vec{w} \cdot \vec{v} - \vec{v} \cdot \vec{w} + \vec{w} \cdot \vec{w}$$
$$= b^2 + c^2 - 2 (\vec{v} \cdot \vec{w}) .$$

We now have

$$b^2 + c^2 - 2bc \cos(A) = a^2 = b^2 + c^2 - 2(\vec{v} \cdot \vec{w}),$$

and we see that $\vec{v} \cdot \vec{w} = bc \cos(A)$. Since $b = |\vec{v}|$ and $c = |\vec{w}|$, we have a geometric interpretation of the dot product of two vectors: It is the product of their lengths times the cosine of the angle between them.

In particular, if $\vec{v}$ and $\vec{w}$ are perpendicular vectors, then since $\cos(90°) = 0$, we see that $\vec{v} \cdot \vec{w} = 0$. Conversely, if $\vec{v}$ and $\vec{w}$ are nonzero, then $|\vec{v}| \neq 0 \neq |\vec{w}|$, and thus if $\vec{v} \cdot \vec{w} = 0$, the only possibility is that $\cos(A) = 0$. Nonzero vectors are perpendicular, therefore, if and only if their dot product is zero.

(5.6) PROBLEM.   Use vector methods to show that the altitudes of $\triangle ABC$ must be concurrent.

Solution.   Let H be the intersection of the altitudes from A and from B so that our task is to show that H also lies on the altitude from C. If H is the point C, there is nothing to prove, and so we can assume H is different from C, and we need to show that line CH is perpendicular to AB. Since the vectors $\overrightarrow{CH}$ and $\overrightarrow{AB}$ are nonzero, it suffices to show that $\overrightarrow{CH} \cdot \overrightarrow{AB} = 0$, and so we want

$$(\vec{H} - \vec{C}) \cdot (\vec{B} - \vec{A}) = 0.$$

Since we do not have a formula that expresses $\vec{H}$ in terms of $\vec{A}$, $\vec{B}$, and $\vec{C}$, we cannot complete the proof simply by plugging and computing, and so we need to be a bit more clever. By the distributive law, the equation we need to prove is equivalent to

$$\vec{H} \cdot (\vec{B} - \vec{A}) = \vec{C} \cdot (\vec{B} - \vec{A}).$$

But we know that H lies on the altitude from A, and from this information, we deduce that $(\vec{H} - \vec{A}) \cdot (\vec{B} - \vec{C}) = 0$, and so

$$\vec{H} \cdot (\vec{B} - \vec{C}) = \vec{A} \cdot (\vec{B} - \vec{C}).$$

Since H also lies on the altitude from B, similar reasoning yields

$$\vec{H} \cdot (\vec{C} - \vec{A}) = \vec{B} \cdot (\vec{C} - \vec{A}).$$

If we add these two equations and use the distributive law a few more times and the commutative law twice, we get

$$\begin{aligned}\vec{H} \cdot (\vec{B} - \vec{A}) &= \vec{A} \cdot (\vec{B} - \vec{C}) + \vec{B} \cdot (\vec{C} - \vec{A}) \\ &= \vec{A} \cdot \vec{B} - \vec{A} \cdot \vec{C} + \vec{B} \cdot \vec{C} - \vec{B} \cdot \vec{A} \\ &= \vec{B} \cdot \vec{C} - \vec{A} \cdot \vec{C} \\ &= \vec{C} \cdot (\vec{B} - \vec{A}),\end{aligned}$$

as desired.   ∎

(5.7) PROBLEM.   Use vectors to show that the circumcenter of $\triangle ABC$ is collinear with the orthocenter and the centroid.

Recall that the circumcenter O, the centroid G, and the orthocenter H actually lie on the Euler line, and the point O lies on the opposite side of G from H, and $HG = 2GO$. We will use this advance knowledge of the relative positions of H, G, and O to find a proof, but we stress that the proof can stand alone.

Solution to Problem 5.7. Since we have the freedom to allow our origin to be any point in the plane, we can take it to be the point that we know will turn out to be the circumcenter. But of course, we do not assume that this point, which we will call O, is the circumcenter; we have to prove that. Following this strategy, we choose the origin as follows. If H and G are the same point, we let O be this point too. Otherwise, we choose O on line HG, on the opposite side of G from H, and half as far from G as H is. Since O is collinear with H and G, it suffices to show that O actually is the circumcenter. We need to show, therefore, that the three distances OA, OB, and OC are all equal.

Because of the way we constructed the origin O, we have $\overrightarrow{OH} = 3\overrightarrow{OG}$. Using our notational shortcut, we can rewrite this equation as $\vec{H} = 3\vec{G} = \vec{A} + \vec{B} + \vec{C}$, where the second equality follows from Theorem 5.2. With this particular choice of origin, therefore, we have $\vec{H} - \vec{A} = \vec{B} + \vec{C}$. Since AH is perpendicular to BC, this yields

$$0 = (\vec{H} - \vec{A}) \cdot (\vec{C} - \vec{B}) = (\vec{B} + \vec{C}) \cdot (\vec{B} - \vec{C}) = \vec{B} \cdot \vec{B} - \vec{C} \cdot \vec{C},$$

and thus $|\vec{B}|^2 = \vec{B} \cdot \vec{B} = \vec{C} \cdot \vec{C} = |\vec{C}|^2$, and we have $|\vec{B}| = |\vec{C}|$. But recall that $\vec{B} = \overrightarrow{OB}$, and hence $|\vec{B}|$ is the distance OB. Similarly, $|\vec{C}| = OC$, and so we have proved that $OB = OC$, and thus O is equidistant from B and C. Similarly, O is equidistant from A and C, and thus O really is the circumcenter of $\triangle ABC$, as desired. ∎

Exercises 5C

5C.1 In Figure 5.4, two squares share vertex O, and line segments AC and BD are drawn connecting vertices of the two squares. The midpoint P of AC is constructed, and line OP is drawn and extended to meet BD at Q. Prove that OP is perpendicular to BD.

HINT: Take the origin to be at O and show that $\vec{A} \cdot \vec{D} = \vec{B} \cdot \vec{C}$. Compute $\vec{P} \cdot (\vec{D} - \vec{B})$.

NOTE: Actually, there is more going on here. We shall see in Exercise 5F.1 that, in fact, $BD = 2OP$.

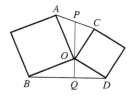

Figure 5.4

5D   Checkerboards

So far, we have done almost nothing with vector techniques of proof that could not easily be done without vectors. (Although one must be clever to find a vector-free proof of Exercise 5C.1, once one finds it, it is not long or difficult.) In this section, and even more in Section F, we will demonstrate how powerful these vector methods of proof really are.

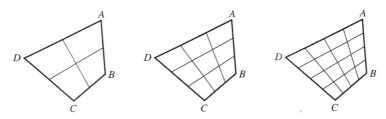

Figure 5.5

Let $ABCD$ be a convex quadrilateral. In other words, all of its angles are less than $180°$. Now divide each side of $ABCD$ into n equal parts, where n is some fixed positive integer, and join the corresponding points, as in Figure 5.5, to form a crisscross pattern that we shall call an $n \times n$ **checkerboard**. (This is definitely not standard nomenclature.) In Figure 5.5, for example, we have drawn 2×2, 3×3, and 4×4 checkerboards, all based on the same quadrilateral.

Consider a 2×2 checkerboard. We know that the midpoints of the four sides of the quadrilateral $ABCD$ are the vertices of a parallelogram by Problem 5.3, for example. The two crossing line segments of the 2×2 checkerboard are the diagonals of this parallelogram, and hence they bisect each other, and thus each of the six line segments that make up a 2×2 checkerboard is cut into two equal pieces. We are referring, of course, to the four sides of the original quadrilateral plus the two crisscross lines. Similarly, but much less obviously, each of the eight line segments that make up a 3×3 checkerboard is divided into three equal pieces. (We will prove this in a moment.) More generally, an $n \times n$ checkerboard is made up of $4 + 2(n - 1)$ line segments, and it turns out that each of these segments is divided into n equal pieces. Of course, by the definition of a checkerboard, we know that each side of the original quadrilateral is divided into n equal pieces; the surprise here is that the $2n - 2$ crisscross segments are also equally divided.

(5.8) THEOREM.   *Each of the $2n + 2$ line segments that comprise an $n \times n$ checkerboard is cut into n equal pieces.*

Proof.   Recall that each of the four sides of quadrilateral $ABCD$ is divided into n equal parts. Point P in Figure 5.6 is one of the division points on side AB, and Q is the corresponding division point on side DC. Then PQ is one of the crisscross line segments, and we have $AP/AB = k/n = DQ/DC$, where k is some integer such that $0 < k < n$. Similarly, R and S are corresponding division points on sides AD

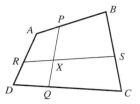

Figure 5.6

and BC, and thus RS is a crisscross line segment and we have $AR/AD = l/n = BS/BN$, where l is an integer with $0 < l < n$.

We need to show that PQ cuts RS at a point that lies exactly k/n of the way from R to S as we move along RS and that this intersection point lies exactly l/n of the way from P to Q along PQ. Of course, we have to prove this for all choices of k and l.

For notational simplicity, let us write $\alpha = k/n$ and $\beta = l/n$. By Lemma 5.5, therefore, we obtain the following vector descriptions of the points P, Q, R, and S:

$$\vec{P} = (1 - \alpha)\vec{A} + \alpha\vec{B}$$
$$\vec{Q} = (1 - \alpha)\vec{D} + \alpha\vec{C}$$
$$\vec{R} = (1 - \beta)\vec{A} + \beta\vec{D}$$
$$\vec{S} = (1 - \beta)\vec{B} + \beta\vec{C}.$$

Now let X be the point on RS that we expect is the point where PQ crosses RS. In other words, X is the point that lies α of the way from R to S along RS. Similarly, let Y be the point on PQ that we expect lies on RS so that Y lies β of the way from P to Q along PQ. Our goal is to show that X and Y are the same point, and we proceed by finding formulas that express $\vec{X}$ and $\vec{Y}$ in terms of the given points A, B, C, and D. As is usual with these vector proofs, we simply need to do some calculations and then compare the results. We expect, of course, to get identical formulas for $\vec{X}$ and for $\vec{Y}$.

We compute

$$\vec{X} = (1 - \alpha)\vec{R} + \alpha\vec{S}$$
$$= (1 - \alpha)\left((1 - \beta)\vec{A} + \beta\vec{D}\right) + \alpha\left((1 - \beta)\vec{B} + \beta\vec{C}\right)$$
$$= (1 - \alpha)(1 - \beta)\vec{A} + \alpha(1 - \beta)\vec{B} + \alpha\beta\vec{C} + (1 - \alpha)\beta\vec{D}.$$

Similarly,

$$\vec{Y} = (1 - \beta)\vec{P} + \beta\vec{Q}$$
$$= (1 - \beta)\left((1 - \alpha)\vec{A} + \alpha\vec{B}\right) + \beta\left((1 - \alpha)\vec{D} + \alpha\vec{C}\right)$$
$$= (1 - \alpha)(1 - \beta)\vec{A} + \alpha(1 - \beta)\vec{B} + \alpha\beta\vec{C} + (1 - \alpha)\beta\vec{D}.$$

Thus $\vec{X} = \vec{Y}$, as we expected, and hence X and Y are the same point. Since this point must be the point of intersection of PQ and RS, the proof is complete. ∎

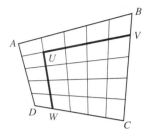

Figure 5.7

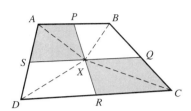

Figure 5.8

Now consider what happens if we remove one row and one column of boxes from a checkerboard. In Figure 5.7, we have drawn a 5×5 checkerboard $ABCD$, and we focus on the smaller quadrilateral $UVCW$, as indicated.

By Theorem 5.8, we know that all of the pieces on each crisscross line of the original checkerboard are equal, and so we see that UV and UW are each divided into four equal pieces. It follows that $UVCW$ is a 4×4 checkerboard. The same thing clearly works in general: We can create an $(n-1) \times (n-1)$ checkerboard from an $n \times n$ checkerboard by deleting the first row and first column of boxes.

Now we come to the amazing part of the theory of checkerboards. To introduce this topic, let us first consider a 2×2 checkerboard.

(5.9) PROBLEM. Let $ABCD$ be a 2×2 checkerboard, as shown in Figure 5.8, where two of the four boxes have been shaded. Show that the shaded area is exactly half of the total area of the checkerboard.

Solution. Let P, Q, R, and S be the midpoints of the sides of quadrilateral $ABCD$, as shown, and let X be the point where PR meets QS. Draw the line segments joining X to A, B, C, and D and note that this partitions the total area of quadrilateral $ABCD$ into four triangular pieces: $\triangle AXB$, $\triangle BXC$, $\triangle CXD$, and $\triangle DXA$. It suffices, therefore, to show that exactly half of the area of each of these four triangles is shaded. But $AP = PB$, and thus $\triangle APX$ and $\triangle BPX$ have equal bases AP and PB, and they have equal altitudes. It follows that $\triangle APX$ and $\triangle BPX$ have equal areas, and this proves that exactly half of the area of $\triangle AXB$ is shaded. A similar argument works for each of the other three triangles that comprise quadrilateral $ABCD$, and it follows that exactly half of the total area of the quadrilateral is shaded.   ∎

Now for the surprise: There is a nice generalization of Problem 5.8 that holds for all $n \times n$ checkerboards and not just in the case where $n = 2$. If we shade the boxes along the diagonal of any $n \times n$ checkerboard, we will prove that the total area of the n shaded boxes is exactly $1/n$ of the area of the entire checkerboard. Of course, we have shaded exactly one nth of the n^2 boxes, but since in general, the boxes do not all have equal areas, this certainly does not show that we have shaded one nth of the area; something more subtle is going on here. Of course, the case $n = 2$ of this fact is exactly Problem 5.9, and the case $n = 1$ is a triviality with no content.

Actually, something even more amazing is true: We need not restrict ourselves to diagonal boxes. If we shade any n of the n^2 boxes, subject only to the condition that no two of the shaded boxes lie in the same row or column, then exactly one nth of the entire area will be shaded. We omit the proof of this, however, because to give a proof would carry us too far from our goal, which is to demonstrate the utility of vectors in geometry proofs.

(5.10) THEOREM. *Suppose that we are given an arbitrary $n \times n$ checkerboard $ABCD$ with area K_{ABCD}. Writing d to denote the total area of the n boxes along the diagonal of this checkerboard, we have $d = \frac{1}{n} K_{ABCD}$.*

Proof.   The theorem certainly holds when $n = 1$, and hence we can assume that $n \geq 2$. Also, we can suppose that the theorem has already been established for all smaller values of n. In particular, therefore, we can assume that the area of the $n - 1$ diagonal boxes of any $(n - 1) \times (n - 1)$ checkerboard is exactly $1/(n - 1)$ of the total area of that checkerboard. We are proceeding by mathematical induction, and the fact that the result holds when n is replaced by $n - 1$ is referred to as the inductive hypothesis.

Let PQ and RS be the leftmost and uppermost of the crisscross lines of the $n \times n$ checkerboard $ABCD$ and let X be the point where these lines meet, as shown in Figure 5.9. Thus quadrilateral $APXR$ is the uppermost of the n diagonal boxes whose total area d we need to compute. We have shaded this box in the figure, and we are to imagine that there are $n - 1$ more shaded boxes, all of which lie inside quadrilateral $XSCQ$. In fact, quadrilateral $XSCQ$ is an $(n - 1) \times (n - 1)$ checkerboard. (Recall that this is a consequence of Theorem 5.8, which was proved using vector methods.) It follows from our inductive hypothesis that the area of the $n - 1$ diagonal boxes inside quadrilateral $XSCQ$ is $\frac{1}{n-1} K_{XSCQ}$, where we are using our standard K notation to indicate the area of a figure. We conclude that the total shaded diagonal area d of the entire checkerboard $ABCD$ is given by the formula

$$d = K_{APXR} + \frac{K_{XSCQ}}{n - 1}.$$

We want to show that $d = \frac{1}{n} K_{ABCD}$, and so we need to prove that $nd = K_{ABCD}$. To accomplish this, we join X to each of the points A, B, C, and D. Since

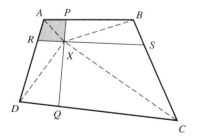

Figure 5.9

$AP = \frac{1}{n}AB$ and $AR = \frac{1}{n}AD$, we see that $K_{AXP} = \frac{1}{n}K_{AXB}$ and $K_{AXR} = \frac{1}{n}K_{AXD}$. Adding these equations and multiplying by n, we obtain

$$nK_{APXR} = K_{ABXD} .$$

Similarly, since $QC = \frac{n-1}{n}QD$ and $SC = \frac{n-1}{n}BC$, we have $K_{XQC} = \frac{n-1}{n}K_{XDC}$ and $K_{XSC} = \frac{n-1}{n}K_{KBC}$. If we add these equations and multiply by n, we get

$$nK_{XSCQ} = (n-1)K_{DXBC} .$$

Now by combining our equations, we get

$$nd = nK_{APXR} + \frac{nK_{XSCQ}}{n-1}$$

$$= K_{ABXD} + \frac{(n-1)K_{DXBC}}{n-1}$$

$$= K_{ABXD} + K_{DXBC}$$

$$= K_{ABCD} ,$$

as required.   ■

Exercises 5D

5D.1 Figure 5.10 shows a 2×3 part of some checkerboard. Show that the sum of the areas of the two boxes labeled 1 is equal to the sum of the areas of the two boxes labeled 2.

NOTE: The corresponding result about the four corner boxes of any part of a checkerboard bounded by two "horizontal" and two "vertical" lines is also valid. This is the key to a proof of the generalization of Theorem 5.10 concerning not-necessarily-diagonal sets of n boxes of an $n \times n$ checkerboard.

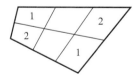

Figure 5.10

5E   A Bit of Trigonometry

In Section F, we shall need to refer to the so-called addition formulas for sine and cosine, and so we digress briefly to review these formulas here.

(5.11) THEOREM.   *The following formulas hold for all angles α and β.*
   a. $\cos(\alpha + \beta) = \cos(\alpha)\cos(\beta) - \sin(\alpha)\sin(\beta)$.
   b. $\sin(\alpha + \beta) = \sin(\alpha)\cos(\beta) + \cos(\alpha)\sin(\beta)$.

The following easy proof uses dot products of vectors and coordinate geometry.

Proof of Theorem 5.11. In the coordinate plane, let O be the origin, let P be the point $(0, 1)$, and let A and B be the points on the unit circle such that $\angle POA = \alpha$ and $\angle POB = \beta$. Since the coordinates of A are $(\cos(\alpha), \sin(\alpha))$ and the coordinates of B are $(\cos(\beta), \sin(\beta))$, we can write $\overrightarrow{OA} = (\cos(\alpha), \sin(\alpha))$ and $\overrightarrow{OB} = (\cos(\beta), \sin(\beta))$.

Recall that we showed that the dot product of two vectors is equal to the product of their lengths times the angle between them. Since $|\overrightarrow{OA}| = OA = 1$ and $|\overrightarrow{OB}| = OB = 1$ and the angle between these vectors is $\alpha - \beta$, we see that $\overrightarrow{OA} \cdot \overrightarrow{OB} = \cos(\alpha - \beta)$. By the definition of the dot product, however, we have $\overrightarrow{OA} \cdot \overrightarrow{OB} = \cos(\alpha)\cos(\beta) + \sin(\alpha)\sin(\beta)$, and thus we conclude that

$$\cos(\alpha - \beta) = \cos(\alpha)\cos(\beta) + \sin(\alpha)\sin(\beta).$$

Addition formula (a) follows from this equation by substituting $-\beta$ for β. (Recall that $\cos(-\beta) = \cos(\beta)$, but $\sin(-\beta) = -\sin(\beta)$.) Finally, to prove (b), we compute that

$$\sin(\alpha + \beta) = \cos(90° - \alpha - \beta)$$
$$= \cos(90° - \alpha)\cos(\beta) + \sin(90° - \alpha)\sin(\beta)$$
$$= \sin(\alpha)\cos(\beta) + \cos(\alpha)\sin(\beta),$$

where the second equality follows by substituting $90° - \alpha$ for α in the equation of the previous paragraph. ∎

There is another more purely geometric proof of the addition formulas that we cannot resist presenting. For brevity, however, we will only consider the case where $0 < \alpha + \beta < 90°$, so that Figure 5.11 applies.

Alternative Proof of Theorem 5.11. In the figure, we started with line OC, and then we drew OB and OA so that $\angle BOC = \alpha$ and $\angle AOB = \beta$. Next, we dropped perpendiculars AW and AV from A to OB and OC, respectively, and then we dropped perpendiculars WX and WU from W to AV and OC, respectively. We see that $\angle WAP$ and $\angle VOP$ are complementary to the equal vertical angles $\angle APW$ and $\angle OPV$, and thus $\angle WAP = \angle VOP = \alpha$, as indicated.

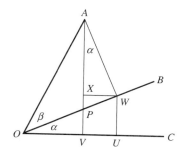

Figure 5.11

We can assume that the length OA is 1 unit, and thus $OV = \cos(\alpha + \beta)$, and we see that $OV = OU - VU$. We have $OW = \cos(\beta)$ and $OU/OW = \cos(\alpha)$, and thus $OU = \cos(\alpha)\cos(\beta)$. Also $AW = \sin(\beta)$ and $WX/AW = \sin(\alpha)$, and this yields $WX = \sin(\alpha)\sin(\beta)$. But $WX = VU$ since $XWUV$ is a rectangle, and hence we have

$$\cos(\alpha + \beta) = OV = OU - VU = OU - WX = \cos(\alpha)\cos(\beta) - \sin(\alpha)\sin(\beta),$$

as required.

Now $WU/OW = \sin(\alpha)$, and thus we have $WU = \sin(\alpha)\cos(\beta)$. Also, $AX/AW = \cos(\alpha)$, and hence $AX = \cos(\alpha)\sin(\beta)$. We now see that

$$\sin(\alpha + \beta) = AV = XV + AX = WU + AX = \sin(\alpha)\cos(\beta) + \cos(\alpha)\sin(\beta),$$

and the proof is complete.   ∎

5F   Linear Operators

Vectors become much more powerful as a technique of proof if we add one more ingredient: linear operators. An **operator** is a function T that yields a vector whenever we plug in a vector. In other words, if $\bar{v}$ is any vector, then $T(\bar{v})$ is some vector determined by $\bar{v}$ according to some specific rule. An easy, but not especially interesting, example of an operator is given by the formula $T(\bar{v}) = -\bar{v}$. In this case, of course, we can think of T as the operation of reversing the direction of all arrows representing vectors, or equivalently: T rotates all arrows by $180°$.

More generally, given any number θ, we can consider the operator T that rotates arrows representing vectors counterclockwise through θ degrees. We should really call this operator something like T_θ so as to emphasize that we have a particular amount of rotation in mind, but we prefer not to clutter our notation with inessential subscripts. Also, we mention that there is nothing especially important about our choice of counterclockwise for the direction of rotation; it is important, however, to be specific in this regard.

It is not difficult to give an explicit formula describing this rotation operator, although we shall not really need to use the formula we are about to derive. If $\bar{v} = (a, b)$, we want to express the coordinates c and d of the vector $T(\bar{v}) = (c, d)$ in terms of the coordinates a and b and the angle of rotation θ. For this purpose, we can suppose that $\bar{v} = \overrightarrow{PQ}$ and $T(\bar{v}) = \overrightarrow{PR}$ so that $\angle QPR = \theta$, as shown in Figure 5.12.

If α is the angle between $\overrightarrow{PQ}$ and the horizontal vector $(0, 1)$, we see that $a = r\cos(\alpha)$ and $b = r\sin(\alpha)$, where $r = |\bar{v}| = PQ$. The angle between $\overrightarrow{PR}$ and the horizontal is $\alpha + \theta$, and the length $PR = PQ = r$. (This is because our rotation operator T does not change the lengths of vectors.) It follows that

$$(c, d) = T(\bar{v}) = \overrightarrow{PR} = (r\cos(\alpha + \theta), r\sin(\alpha + \theta)),$$

Figure 5.12

and thus by Theorem 5.11, we get formulas for the coordinates of $T(\bar{v})$:

$$c = r\cos(\alpha + \theta) = r(\cos(\alpha)\cos(\theta) - \sin(\alpha)\sin(\theta)) = a\cos(\theta) - b\sin(\theta)$$
$$d = r\sin(\alpha + \theta) = r(\cos(\alpha)\sin(\theta) + \sin(\alpha)\cos(\theta)) = a\sin(\theta) + b\cos(\theta).$$

For the benefit of readers familiar with matrix notation, we can write these transformation equations as

$$(c, d) = (a, b)\begin{pmatrix} \cos(\theta) & \sin(\theta) \\ -\sin(\theta) & \cos(\theta) \end{pmatrix},$$

and thus $T(\bar{v}) = \bar{v}A$, where A is the 2×2 matrix of sines and cosines in the preceding formula. In other words, the operator T simply multiplies a vector (on the right) by the matrix A.

We shall show that the operator T that rotates all vectors counterclockwise through a given angle θ is, in fact, a linear operator, which means that the operator respects both addition and scalar multiplication of vectors. More precisely, we say that an operator T is **linear** if $T(\bar{v} + \bar{w}) = T(\bar{v}) + T(\bar{w})$ and $T(z\bar{v}) = zT(\bar{v})$ for all vectors $\bar{v}$ and $\bar{w}$ and for all scalars z.

To check that our rotation operator is linear, we choose arbitrary vectors $\bar{v}$ and $\bar{w}$, and we compute that

$$T(\bar{v} + \bar{w}) = (\bar{v} + \bar{w})A = \bar{v}A + \bar{w}A = T(\bar{v}) + T(\bar{w}),$$

as required. Here, A is the matrix of sines and cosines, as earlier, and we are using the distributive law of matrix multiplication. Also, if z and $\bar{v}$ are an arbitrary scalar and an arbitrary vector, we have

$$T(z\bar{v}) = (z\bar{v})A = z(\bar{v}A) = zT(\bar{v}),$$

as we wanted. These two simple calculations show that the rotation operator T really is a linear operator, as we asserted.

It is also possible to see geometrically why the rotation operator T is linear. To show that T respects addition, it is convenient to think of the geometric significance of vector addition in a way that is slightly different from our previous approach. Suppose that we represent the vectors $\bar{v}$, $\bar{w}$, and their sum $\bar{v} + \bar{w}$ as arrows all having the same tail P. If $\bar{v} = \overrightarrow{PQ}$ and $\bar{w} = \overrightarrow{PR}$ and we write $\bar{v} + \bar{w} = \overrightarrow{PS}$, we ask how the point S is determined from a knowledge of the three points P, Q, and R.

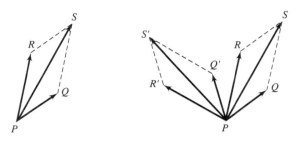

Figure 5.13

Consider the left diagram in Figure 5.13. We drew a line through Q parallel to PR and a line through R parallel to PQ, and we let S be the point where these two lines meet. Then $PQSR$ is a parallelogram, and hence we have $\overrightarrow{PR} = \overrightarrow{QS}$. We see now that $\overrightarrow{PQ} + \overrightarrow{PR} = \overrightarrow{PQ} + \overrightarrow{QS} = \overrightarrow{PS}$. This shows how to add two vectors represented by arrows with a common tail: It suffices to complete the parallelogram. The arrow (with the same tail) along the diagonal of the parallelogram represents the sum of the two original vectors.

In the right diagram of Figure 5.13, we drew another copy of parallelogram $PQSR$, and then we rotated this entire figure counterclockwise through θ degrees about point P. The result of this rotation is parallelogram $PQ'S'R'$, and it should be clear that $T(\overrightarrow{PQ}) = \overrightarrow{PQ'}$, $T(\overrightarrow{PR}) = \overrightarrow{PR'}$, and $T(\overrightarrow{PS}) = \overrightarrow{PS'}$. We can now see that

$$T(\overrightarrow{PQ} + \overrightarrow{PR}) = T(\overrightarrow{PS}) = \overrightarrow{PS'} = \overrightarrow{PQ'} + \overrightarrow{PR'} = T(\overrightarrow{PQ}) + T(\overrightarrow{PR}),$$

and thus the operator T respects vector addition. (Recall that this is one of the two things we must show to establish that an operator is linear.) To see that the rotation operator T also respects scalar multiplication, and hence is linear, it is enough to observe that if we stretch a vector and then rotate it, the result is the same as would have been obtained had we rotated the vector first and then stretched it.

We can now demonstrate how powerful linear operators are as a technique of geometric proof. In the following, the point we refer to as the center of a square is the unique point that is equidistant from the four vertices. This point, of course, is also the intersection of the diagonals of the square.

(5.12) PROBLEM.   Outward-facing squares are drawn on the sides of an arbitrary quadrilateral $ABCD$, as shown in Figure 5.14. If P, Q, R, and S are the centers of these four squares, as shown, prove that line segments PR and SQ are equal and perpendicular.

Solution.   Let T be the linear operator corresponding to a 90° counterclockwise rotation. If we can show that $T(\overrightarrow{PR}) = \overrightarrow{SQ}$, it will follow that $PR = SQ$ and also that PR is perpendicular to SQ, and that will complete the proof. To decide exactly what vector equation we need to prove, we must examine the diagram carefully so as to avoid confusing counterclockwise with clockwise. In this problem, we definitely must show that $T(\overrightarrow{PR})$ equals $\overrightarrow{SQ}$; the proof would not work if we carelessly replaced $\overrightarrow{SQ}$ with $\overrightarrow{QS}$.

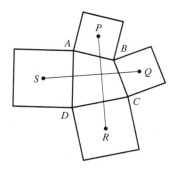

The independent variables in this problem are the four vertices A, B, C, and D of the given quadrilateral, because once those points are known, it is clear that the other relevant points P, Q, R, and S are unambiguously determined. We attempt, therefore, to express $\vec{P}$, $\vec{Q}$, $\vec{R}$, and $\vec{S}$ in terms of $\vec{A}$, $\vec{B}$, $\vec{C}$, and $\vec{D}$.

We begin by finding a formula for $\vec{P}$ in terms of $\vec{A}$ and $\vec{B}$. To do this, we let U be the midpoint of AB, and we note that $PU = AU$ and PU is perpendicular to AU. (These

Figure 5.14

observations are immediate consequences of the fact that P is the center of the square with side AB.) It follows that $T(\overrightarrow{AU}) = \overrightarrow{UP}$, and thus, using the linearity of the operator T, we have

$$T(\vec{U}) - T(\vec{A}) = T(\vec{U} - \vec{A}) = T(\overrightarrow{AU}) = \overrightarrow{UP} = \vec{P} - \vec{U},$$

and this yields $\vec{P} = \vec{U} + T(\vec{U}) - T(\vec{A})$. Since U is the midpoint of AB, we know that $\vec{U} = \frac{1}{2}(\vec{A} + \vec{B})$, and we can substitute this into our formula for $\vec{P}$. Using the linearity of T again, we obtain

$$\vec{P} = \frac{1}{2}\left(\vec{A} + \vec{B} + T(\vec{A}) + T(\vec{B})\right) - T(\vec{A})$$

$$= \frac{1}{2}\left(\vec{A} + \vec{B} - T(\vec{A}) + T(\vec{B})\right).$$

Now we must find similar expressions for $\vec{Q}$, $\vec{R}$, and $\vec{S}$. As usual, we can do this with almost no work; we simply march around the quadrilateral, replacing A by B, B by C, C by D, and D by A. We repeat our formula for $\vec{P}$ and modify it to get the three other formulas we need:

$$\vec{P} = \frac{1}{2}\left(\vec{A} + \vec{B} - T(\vec{A}) + T(\vec{B})\right)$$

$$\vec{Q} = \frac{1}{2}\left(\vec{B} + \vec{C} - T(\vec{B}) + T(\vec{C})\right)$$

$$\vec{R} = \frac{1}{2}\left(\vec{C} + \vec{D} - T(\vec{C}) + T(\vec{D})\right)$$

$$\vec{S} = \frac{1}{2}\left(\vec{D} + \vec{A} - T(\vec{D}) + T(\vec{A})\right).$$

Next, we compute

$$\overrightarrow{PR} = \vec{R} - \vec{P}$$

$$= \frac{1}{2}\left(\vec{C} + \vec{D} - T(\vec{C}) + T(\vec{D}) - \vec{A} - \vec{B} + T(\vec{A}) - T(\vec{B})\right) .$$

We need to compute $T(\overrightarrow{PR})$, and so we have to apply the linear operator T to the right side of this equation. For that purpose, we need to know how to compute $T(T(\vec{v}))$, where $\vec{v}$ is an arbitrary vector. But this is easy. Since T is a $90°$ rotation, we see that applying it twice yields a $180°$ rotation, and it follows that $T(T(\vec{v})) = -\vec{v}$. Using this fact, together with the linearity of T, we obtain

$$T(\overrightarrow{PR}) = \frac{1}{2}\left(T(\vec{C}) + T(\vec{D}) + \vec{C} - \vec{D} - T(\vec{A}) - T(\vec{B}) - \vec{A} + \vec{B}\right) .$$

Finally, we observe that

$$\overrightarrow{SQ} = \vec{Q} - \vec{S}$$

$$= \frac{1}{2}\left(\vec{B} + \vec{C} - T(\vec{B}) + T(\vec{C}) - \vec{D} - \vec{A} + T(\vec{D}) - T(\vec{A})\right) ,$$

and this is identical, except for a rearrangement of the terms, with our formula for $T(\overrightarrow{PR})$. We conclude that $T(\overrightarrow{PR}) = \overrightarrow{SQ}$, and thus line segments PR and SQ are equal and perpendicular, as desired.   ∎

The following is another example of what would be a difficult problem if one had to rely on cleverness. It can be done by a mere computation, however, using vectors and linear operators.

(5.13) PROBLEM.   In Figure 5.15, equilateral triangles $\triangle PAB$, $\triangle PCD$, and $\triangle PEF$ share a vertex. The remaining six vertices of these three triangles are joined in pairs by line segments FA, BC, and DE, as shown, and points X, Y, and Z are the midpoints of these three segments. Prove that $\triangle XYZ$ is equilateral.

Solution.   Because both the given data and the desired conclusion involve equilateral triangles, it should be clear that the linear operator that rotates vectors counterclockwise through $60°$ is relevant, and we call this operator T. Also, it is convenient, but

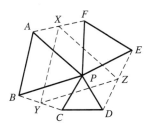

Figure 5.15

certainly not necessary, to choose our origin at P so that, for example, we can write $\overrightarrow{PA}$ simply as $\vec{A}$. Thus $\vec{B} = T(\vec{A})$, $\vec{D} = T(\vec{C})$, and $\vec{F} = T(\vec{E})$. Since we can view the points A, C, and E as the independent variables in this problem, our strategy will be to express $\vec{X}$, $\vec{Y}$, and $\vec{Z}$ in terms of $\vec{A}$, $\vec{C}$, and $\vec{E}$. To complete the argument, it will suffice to prove that $T(\overrightarrow{ZX}) = \overrightarrow{ZY}$ since that will show that $\triangle XYZ$ is isosceles with base XY and vertex angle equal to $60°$. It follows that each base angle is thus equal to $60°$, and so $\triangle XYZ$ is equiangular, and hence is equilateral, as desired.

As is usual with this type of proof, this method *must* work. If, when we compute $T(\overrightarrow{ZY})$, it turns out to be unequal to $\overrightarrow{ZX}$, this shows that we have made an error in computation or in setting up the equations.

Since X is the midpoint of AF, we can write $\vec{X} = \frac{1}{2}(\vec{A} + \vec{F})$. But since $\vec{F} = T(\vec{E})$, we can substitute to obtain

$$\vec{X} = \frac{1}{2}\left(\vec{A} + T(\vec{E})\right) .$$

As usual, we can march around the figure to obtain the corresponding formulas:

$$\vec{Y} = \frac{1}{2}\left(\vec{C} + T(\vec{A})\right)$$

$$\vec{Z} = \frac{1}{2}\left(\vec{E} + T(\vec{C})\right) .$$

It follows that

$$\overrightarrow{ZX} = \vec{X} - \vec{Z} = \frac{1}{2}\left(\vec{A} + T(\vec{E}) - \vec{E} - T(\vec{C})\right) ,$$

and we have to apply T to this and check that the result is equal to

$$\overrightarrow{ZY} = \vec{Y} - \vec{Z} = \frac{1}{2}\left(\vec{C} + T(\vec{A}) - \vec{E} - T(\vec{C})\right) .$$

To accomplish this, it is useful to have a formula for $T(T(\vec{v}))$, where $\vec{v}$ is an arbitrary vector. To obtain the correct formula, consider Figure 5.16. In this figure, we assume that $\overrightarrow{PQ} = \vec{v}$, $\overrightarrow{PR} = T(\vec{v})$, and $\overrightarrow{PS} = T(T(\vec{v}))$. Then $PR = PS$ and $\angle RPS = 60°$, and it follows that $\triangle RPS$ is equilateral. Thus $RS = RP = PQ$, and $\angle SRP = 60° = \angle RPQ$. We conclude that RS is parallel and equal to PQ, and hence $\overrightarrow{RS} = -\overrightarrow{PQ} = -\vec{v}$. We thus have

$$T(T(\vec{v})) = T\left(T(\overrightarrow{PQ})\right) = \overrightarrow{PS} = \overrightarrow{PR} + \overrightarrow{RS} = T(\vec{v}) - \vec{v} .$$

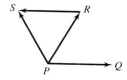

Figure 5.16

We can now compute that

$$T(\vec{ZX}) = \frac{1}{2}T\left(\vec{A} + T(\vec{E}) - \vec{E} - T(\vec{C})\right)$$

$$= \frac{1}{2}\left(T(\vec{A}) + T(T(\vec{E})) - T(\vec{E}) - T(T(\vec{C}))\right)$$

$$= \frac{1}{2}\left(T(\vec{A}) + T(\vec{E}) - \vec{E} - T(\vec{E}) - T(\vec{C}) + \vec{C}\right)$$

$$= \frac{1}{2}\left(T(\vec{A}) - \vec{E} - T(\vec{C}) + \vec{C}\right),$$

and indeed this does agree with our formula for $\vec{ZY}$. The proof is now complete. ■

The following theorem, which is sometimes attributed to Emperor Napoleon Bonaparte, also involves equilateral triangles, and here too, our vector proof depends on the 60° rotation linear operator.

(5.14) THEOREM. *If we construct outward-pointing equilateral triangles on the three sides of an arbitrary given triangle, then the triangle formed by the centroids of the three equilateral triangles is equilateral.*

Proof. In Figure 5.17, we are given $\triangle ABC$ and we have constructed equilateral $\triangle BCP$, $\triangle CAQ$, and $\triangle ABR$, with centroids X, Y, and Z, respectively. Our strategy will be to express $\vec{X}$, $\vec{Y}$, and $\vec{Z}$ in terms of the given data, which are $\vec{A}$, $\vec{B}$, and $\vec{C}$. We will then compute that $T(\vec{XY}) = \vec{XZ}$, where T is the operator that rotates vectors 60° counterclockwise. It will follow that $\triangle XYZ$ is equilateral, as required. Again, we know that this method of proof must work; all that is required is to carry out the computations without error.

Since $\triangle BCP$ is equilateral, we see that $T(\vec{CB}) = \vec{CP}$, and thus $\vec{P} - \vec{C} = T(\vec{B} - \vec{C})$. The linearity of T together with a bit of algebra yields the equation

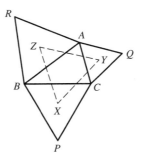

Figure 5.17

$\vec{P} = \vec{C} - T(\vec{C}) + T(\vec{B})$, and since we know by Theorem 5.2 that $\vec{X} = \frac{1}{3}(\vec{B} + \vec{C} + \vec{P})$, we get

$$\vec{X} = \frac{1}{3}\left(\vec{B} + 2\vec{C} - T(\vec{C}) + T(\vec{B})\right).$$

As usual, we can avoid work by marching around the diagram, and we obtain

$$\vec{Y} = \frac{1}{3}\left(\vec{C} + 2\vec{A} - T(\vec{A}) + T(\vec{C})\right) \qquad \text{and}$$

$$\vec{Z} = \frac{1}{3}\left(\vec{A} + 2\vec{B} - T(\vec{B}) + T(\vec{A})\right).$$

These equations yield

$$\overrightarrow{XY} = \vec{Y} - \vec{X}$$

$$= \frac{1}{3}\left(\vec{C} + 2\vec{A} - T(\vec{A}) + T(\vec{C}) - \vec{B} - 2\vec{C} + T(\vec{C}) - T(\vec{B})\right)$$

$$= \frac{1}{3}\left(2\vec{A} - \vec{B} - \vec{C} - T(\vec{A}) - T(\vec{B}) + 2T(\vec{C})\right),$$

and

$$\overrightarrow{XZ} = \vec{Z} - \vec{X}$$

$$= \frac{1}{3}\left(\vec{A} + 2\vec{B} - T(\vec{B}) + T(\vec{A}) - \vec{B} - 2\vec{C} + T(\vec{C}) - T(\vec{B})\right)$$

$$= \frac{1}{3}\left(\vec{A} + \vec{B} - 2\vec{C} + T(\vec{A}) - 2T(\vec{B}) + T(\vec{C})\right).$$

To compute $T(\overrightarrow{XY})$, we will use the fact that we derived previously: $T(T(\vec{v})) = T(\vec{v}) - \vec{v}$ for an arbitrary vector $\vec{v}$. We have

$$T(\overrightarrow{XY}) = \frac{1}{3}\left(2T(\vec{A}) - T(\vec{B}) - T(\vec{C}) - T(\vec{A})\right.$$

$$\left. + \vec{A} - T(\vec{B}) + \vec{B} + 2T(\vec{C}) - 2\vec{C}\right)$$

$$= \frac{1}{3}\left(\vec{A} + \vec{B} - 2\vec{C} + T(\vec{A}) - 2T(\vec{B}) + T(\vec{C})\right).$$

We see that, as expected, we obtained exactly the same formulas for $T(\overrightarrow{XY})$ and $\overrightarrow{XZ}$, and this completes the proof. ∎

Since equilateral triangles appear both in the statement and in the conclusion of Napoleon's theorem (Theorem 5.14), and since a special property of the 60° rotation operator was used in the proof, it seems surprising that, in fact, there is a generalization of Napoleon's theorem that has nothing to do with 60° angles or with equilateral triangles.

(5.15) THEOREM.   *Suppose that three similar outward-pointing triangles are con-structed on the sides of an arbitrary $\triangle ABC$, as shown in Figure 5.18, where $\triangle PCB \sim \triangle CQA \sim \triangle BAR$. If X, Y, and Z are, respectively, the centroids of these three similar triangles, then $\triangle XYZ$ is similar to each of them.*

Of course, if the three similar triangles on the sides of $\triangle ABC$ happen to be equilateral, then Theorem 5.15 tells us that $\triangle XYZ$ is also equilateral, and thus we see that Napoleon's theorem is included in Theorem 5.15. We close this chapter with a proof of this result that uses vector techniques and linear operators, but which does not seem to be quite so mechanical or lacking in cleverness as are our other vector proofs.

Proof of Theorem 5.15.   Our first task is to define an appropriate linear operator T. We would like to have $T(\overrightarrow{PC}) = \overrightarrow{PB}$, $T(\overrightarrow{CQ}) = \overrightarrow{CA}$, and $T(\overrightarrow{BA}) = \overrightarrow{BR}$. Although it may seem that this is asking for too much, we shall see that because $\triangle PCB \sim \triangle CQA \sim \triangle BAR$, it is possible to define T so that its effect on $\overrightarrow{PC}$, $\overrightarrow{CQ}$, and $\overrightarrow{BA}$ is as desired.

Note that $\angle CPB = \angle QCA = \angle ABR$, and so we want our linear operator T to rotate vectors counterclockwise by this angle, which we call θ. But we do not want T to be a pure rotation; it should also stretch or shrink vectors appropriately. We define T, therefore, to be the result of a counterclockwise rotation through θ followed by multiplication by the scalar $z = PB/PC$. We know that a rotation operator is linear, and it follows easily that our operator T defined by a rotation through some fixed angle followed by multiplication by some fixed scalar is also linear. By the definition of T, we see that $T(\overrightarrow{PC}) = \overrightarrow{PB}$, as wanted. But $\triangle PCB \sim \triangle CQA$, and thus $PC/CQ = PB/CA$, and we see that $z = PB/PC = CA/CQ$. It follows that if we rotate the vector $\overrightarrow{CQ}$ counterclockwise through the angle θ and then multiply by the scalar z, the result is the vector $\overrightarrow{CA}$. Thus $T(\overrightarrow{CQ}) = \overrightarrow{CA}$, and similarly, since we also have $z = BR/BA$, we see that $T(\overrightarrow{BA}) = \overrightarrow{BR}$, as desired.

Our goal will be to show that $T(\overrightarrow{XY}) = \overrightarrow{XZ}$. When we establish this, it will follow that $\angle YXZ = \theta = \angle CPB$. We shall also know that $XZ/XY = z = PB/PC$, and hence $PC/XY = PB/XZ$, and thus $\triangle XYZ \sim \triangle PCB$ by the SAS similarity criterion.

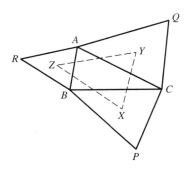

Figure 5.18

We have the following equations:

$$T(\vec{C}) - T(\vec{P}) = T(\overrightarrow{PC}) = \overrightarrow{PB} = \vec{B} - \vec{P}$$
$$T(\vec{Q}) - T(\vec{C}) = T(\overrightarrow{CQ}) = \overrightarrow{CA} = \vec{A} - \vec{C}$$
$$T(\vec{A}) - T(\vec{B}) = T(\overrightarrow{BA}) = \overrightarrow{BR} = \vec{R} - \vec{B},$$

Since $3\vec{Y} = \vec{C} + \vec{Q} + \vec{A}$, $3\vec{X} = \vec{P} + \vec{C} + \vec{B}$, and $3\vec{Z} = \vec{B} + \vec{A} + \vec{R}$, we see that by adding the preceding three equations and multiplying by $1/3$, we obtain

$$T(\vec{Y}) - T(\vec{X}) = \vec{Z} - \vec{X},$$

and thus $T(\overrightarrow{XY}) = \overrightarrow{XZ}$, as desired.   ■

Exercises 5F

5F.1 In the situation of Exercise 5C.1, use an appropriate linear operator to show that BD is perpendicular to OP and that $BD = 2OP$.

5F.2 In the situation of Problem 5.12, assume that quadrilateral $ABCD$ is a parallelogram. Show that $PQRS$ is a square.

CHAPTER SIX

Geometric Constructions

6A Rules of the Game

What information are we allowed to use when we are trying to determine whether or not some statement about triangles or circles is true? Are we permitted to use everything that we know to be correct? Obviously, that depends on our goal. If we simply want to decide whether or not the assertion is true, it would certainly be permissible to consult an advanced geometry text or to ask an expert whom we trust. Even if we are unable to find the specific assertion in the book or if the expert happens to be unfamiliar with it, we might nevertheless learn some other relevant facts from these sources, and we might then be able to use these to prove or disprove our original statement. Directly or indirectly, therefore, books and expert knowledge can often be used to determine the truth or falsity of some given assertion.

But suppose that the assertion whose truth we are trying to establish is one of the exercises in this book. Or suppose that we are teaching a geometry class, and we are trying to demonstrate the deductive method. In such circumstances, it clearly would violate the rules of the game to "prove" something by an appeal to authority because, as we know, those rules permit us to use only previously established theorems or explicitly stated axioms. We stress that the prohibition of appeals to authority is not based on a fear that such appeals are likely to yield wrong answers; appeals to authority are excluded because they simply have no place in the game of deductive mathematics.

The situation is somewhat analogous when we consider what tools we are permitted to use to draw geometric figures. If the purpose of the drawing is simply to obtain an accurate diagram, there are a number of effective tools that could be used. The diagrams in this book were drawn with the aid of a computer, for example, but older drawing tools include rulers, compasses, protractors, and various other devices designed to draw particular angles or line segments with particular lengths. The classical Greek geometers, however, established certain rules for the game of geometric constructions. As we will explain, these Greek rules limit us to just two tools: a straightedge for drawing lines and a compass for drawing circles. The rules also require that we use these tools in certain specified ways. Other tools are not necessarily less accurate; they are excluded simply because they violate the rules of the game.

The only "legal" use for a straightedge is to draw the line determined by two points that have been previously marked on our paper. (In practice, of course, we can draw only a segment of that line.) The official "regulation" straightedge is not a ruler; it has no distance markings on it, and furthermore, we are not allowed to make any marks on it. We shall see that if we are allowed to mark our straightedge, then it is possible to do certain constructions that would otherwise be impossible. Also, we mention that while most rulers have two parallel straight edges, the official straightedge has a single usable edge. With a straightedge alone, therefore, we cannot draw two parallel lines, although as we shall see, it is easy to draw parallel lines using both a straightedge and a compass.

We mention a further prohibited use of a straightedge: drawing tangents to circles. Suppose, for example, that a circle has been drawn on our paper and that a point P is marked on the paper outside of the circle. If we wish to draw one of the two lines through P that are tangent to the circle, it is tempting to place the straightedge so that it runs through the point P and then to rotate it slowly about P until it just touches the circle. This will work, of course, but it is not allowed by the rules of the game. (As we shall see, it is not hard to describe a legal construction of the tangents to a circle from an outside point if we use both a straightedge and a compass.) Similarly, if we wish to draw a tangent to a given circle through a given point Q on the circle, it is not permissible simply to place the straightedge so that it touches the circle at the point Q only. As an alternative, we might try to choose a second point R on the circle, near Q. We certainly can draw the secant line QR, and of course, we know that the desired tangent line is the limit of this secant line as R approaches Q. But it is not legal to draw the tangent line through Q by taking limits. The rules require that a construction must have only finitely many steps, and so we cannot let R get arbitrarily close to Q.

Despite the rules that limit the ways a straightedge can be used, there are a few things we can do with this ideal tool that would be impractical with ordinary mundane drawing instruments. With an official straightedge, for example, we can draw the line determined by any two different points no matter how far apart they are or how close together. With an actual ruler, of course, we could not draw the line joining two points that are farther apart than the length of the ruler, and it would be extremely difficult to draw accurately the line joining two points that are very close together.

The official Greek compass does the following and nothing else: Given two distinct points P and Q on our paper, it draws the unique circle centered at P and passing through Q, or it draws an appropriate arc of that circle. The only requirement is that the points P and Q should be different; it does not matter how close together or far apart they are.

An actual physical compass, of course, is a device that has a hinged pair of arms whose tips are held at a fixed but adjustable distance r apart. One of these tips is sharp and is usually made of metal, and the other is a pen or pencil point. In use, the metal tip punctures the paper, and so it is held stationary at some specified point P, while the pencil draws a circle or an arc centered at P and of radius r. To use an actual compass to draw the circle centered at P and passing through Q, the first step is to place the metal tip at P. Next, the separation between the metal tip and the pencil point is carefully adjusted so that the pencil point just touches Q, and finally, after this adjustment is made, the circle can be drawn.

Suppose that we are given three points P, Q, and R on our paper, and we wish to draw the circle centered at R that has radius equal to PQ. With a physical compass, this is easy. First, place the metal tip at P and adjust the opening of the compass so that the pencil point is at Q. Next, lift the compass from the paper and then, without changing the distance between the metal tip and the pencil point, place the metal tip at R and draw the circle. This procedure is not legal, however, according to the official Greek construction rules. Recall that, officially, all we can do with a compass is draw a circle with a specified center and going through a specified point. One way to think of this is to imagine that an actual physical compass has a memory, whereas the official Greek compass will forget the distance between the metal tip and the pencil point as soon as the metal tip loses contact with the paper. To use a Greek compass to draw the circle with radius PQ centered at R, we first need to construct some point S such that $RS = PQ$, and then we could draw the circle centered at R and running through S. To construct such a point S using only a regulation compass and a regulation straightedge, it is convenient to be able to construct parallel lines.

(6.1) PROBLEM.   Given a line m and a point P not on the line, construct the line through P that is parallel to m.

Solution.   Choose a point A on line m and draw an arc through P, centered at A and meeting m at point B, as shown in Figure 6.1. Next, draw arcs through A centered at P and at B and let Q be the point other than A where the corresponding two circles meet. Then line PQ is the desired parallel to m through P.

   We need to prove that $PQ \| m$, and for this purpose, we observe that $BQ = BA = PA = PQ$ by construction, and thus quadrilateral $APQB$ is a parallelogram because it has two pairs of equal opposite sides. (In fact, $APQB$ is a rhombus, but this is irrelevant.) It follows that $PQ \| AB$, and since line AB is our original line m, we are done.   ∎

We can now show how to do with a straightedge and Greek compass what we can do with an ordinary compass that remembers its setting when its metal tip is lifted from the paper.

(6.2) PROBLEM.   Given distinct points P, Q, and R, construct the circle centered at R and having radius PQ.

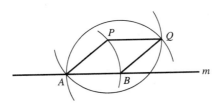

Figure 6.1

Solution.   First, assume that P, Q, and R are not collinear. Draw lines PQ and PR and then, using Problem 6.1, construct the line through Q parallel to PR and the line through R parallel to PQ and let S be the point where these lines meet. Then $QSRP$ is a parallelogram, and so $PQ = RS$. We can now draw the circle through S centered at R, and we see that its radius is equal to PQ, as required.

   Since the actual point Q is irrelevant here, and only the distance PQ is significant, we can replace Q by any other convenient point Q' on the circle through Q centered at P. In particular, in the case where the given points P, Q, and R happen to be collinear, we can replace Q by another point Q' on this circle, where Q' is not collinear with P and R. We can then carry out the construction of the previous paragraph.   ■

We see now that it really does not matter that an official Greek compass does not remember its distance setting when it is lifted from the paper. Using the construction of Problem 6.2, we can pretend whenever it is convenient to do so that our compass does remember its setting, and so henceforth we will change the rules of our game and allow a compass with a memory.

To demonstrate some of what we can do with such an enhanced, powerful compass, we present a few easy but useful constructions.

(6.3) PROBLEM.   Construct the perpendicular bisector of a given line segment.

Solution.   Choose an arbitrary point P on the given segment AB, closer to B than to A, and draw an arc through P centered at A, as in Figure 6.2. Now, using Problem 6.2 to pretend that our compass has a memory, draw an arc of radius AP centered at B, and let X and Y be the two points where these two arcs cross. We claim that line XY is the desired perpendicular bisector.

   To see why, observe that $AX = AP = BX$ and $AY = AP = BY$, and thus each of X and Y is equidistant from both A and B. It follows by Theorem 1.10 that each of X and Y lies on the perpendicular bisector of segment AB, and thus the line through X and Y is, in fact, the perpendicular bisector.   ■

Once we have constructed the perpendicular bisector of a line segment, we can get the midpoint of the segment for free: It is just the intersection of the perpendicular bisector with the original segment. Since we can construct midpoints of line segments, we clearly can construct the medians of a given triangle, and thus we can construct the

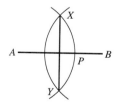

Figure 6.2

centroid of the triangle, which is the intersection of the medians. Of course, we need to construct just two of the three medians to find their intersection. Also, we can construct the circumcenter O of a given triangle since O is the intersection of the perpendicular bisectors of the sides. And of course, it is enough to construct just two of these three perpendicular bisectors. Once we have the circumcenter, it is easy to construct the circumcircle: Just draw the circle centered at O through any one of the vertices of the triangle.

It is also easy to drop a perpendicular from a point to a line, and so we can construct the altitudes of a given triangle, and thus we can construct the orthocenter.

(6.4) PROBLEM.   Given a point P and a line m, construct the line through P that is perpendicular to m.

Note that in Problem 6.4, point P may or may not lie on line m. Although the construction is essentially the same in these two cases, the language that is customarily used is slightly different. If point P does not lie on line m, we "drop" the perpendicular from P to m, but if P does lie on m, we "erect" the perpendicular to m at P.

Solution to Problem 6.4.   First, we construct a pair of points A and B on m such that A and B are equidistant from P. This can be done by drawing a circle centered at P, with a radius large enough so that the circle meets m in two points. By Problem 6.3, we can construct the perpendicular bisector b of line segment AB. Then b is perpendicular to line $m = AB$, and b goes through P, as required, since P is equidistant from A and B.   ■

(6.5) PROBLEM.   Construct the bisector of a given angle.

Solution.   Given $\angle ABC$, as shown in Figure 6.3, construct points P and Q on AB and AC such that $BP = BQ$. Do this by drawing any circle centered at B and letting P and Q be the points where this circle meets the sides of the angle. Next, construct a point R, equidistant from P and Q and different from B. An easy way to do this is to draw the circle through P centered at Q and the circle through Q centered at P and to let R be one of the points of intersection of these two circles. Then $PR = PQ = QR$, as desired.

Line BR is the desired angle bisector. To see why this is true, note that $\triangle BPR \cong \triangle BQR$ by SSS, and thus $\angle PBR = \angle QBR$, as required.   ■

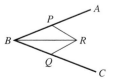

Figure 6.3

Since the incenter of a triangle is the intersection of the three angle bisectors, we now see that it is possible to construct the incenter I of a given $\triangle ABC$. Of course, to find I, it suffices to construct just two of the angle bisectors. Once we have the center of the inscribed circle, how can we construct the circle itself? It clearly suffices to construct any one point that we know must lie on the circle.

Consider the point P of tangency of the inscribed circle of $\triangle ABC$ with side AB. We know that radius IP is perpendicular to tangent AB, and it follows that we can construct P by dropping the perpendicular from I to AB. Point P is, of course, the foot of this perpendicular: the point where the perpendicular line meets AB. Once we have constructed P, we can draw the circle through P centered at I, and that will be the inscribed circle of $\triangle ABC$.

Exercises 6A

6A.1 Given a line segment, construct an equilateral triangle whose sides have lengths equal to the length of the given segment.

6A.2 Given a line segment, construct a square whose sides have lengths equal to the length of the given segment.

6A.3 Given a circle, construct its center.

6A.4 Given two line segments, construct a rhombus whose diagonals have lengths equal to the lengths of the two given segments.

6A.5 Given $\triangle ABC$, where $\angle C$ is obtuse, construct $\triangle PBC$ having the same area, where $\angle P = 90°$.

HINT: Construct a circle with diameter AB.

6A.6 Draw a circle centered at P and choose a point A on the circle. Next, draw the circle with the same radius centered at A and let B be a point where this circle meets the original circle. Now draw the circle with the same radius and centered at B and let C be the point other than A where this circle meets the original circle. Draw the circle with the same radius centered at C and let D be the point other than B where this circle meets the original circle. Continuing like this, construct points E, F, and G. Prove that G is the same point as A and show that hexagon $ABCDEF$ is regular.

6B   Reconstructing Triangles

Given a triangle, we have seen how to construct its medians, altitudes, and angle bisectors. A more difficult and interesting type of problem is to reverse this procedure: Given some data about a triangle, but without being given the triangle itself, we want to reconstruct the original triangle. For example, we might want to reconstruct $\triangle ABC$ if we are given the lengths $b = AC$ and $c = AB$ of two sides of the triangle, and we are also given the length m of median AM, where M is the midpoint of side BC. Of course, the best we can hope for is to construct a triangle *congruent* to the original triangle; there is obviously no way to reconstruct the actual $\triangle ABC$ from the given lengths.

Before we proceed to solve this and similar problems, we should clarify what it means to be given the three lengths a, b, and m. What we are actually given is three line segments drawn on a piece of paper, and we are told that the lengths of these segments are a, b, and m. We are also told which segment has which length.

(6.6) PROBLEM.   Reconstruct a triangle given the lengths of two of its sides and the median to the third side.

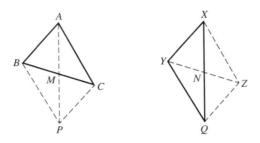

Figure 6.4

Solution.   To explain our construction, we pretend that the original (hidden) triangle is $\triangle ABC$, in the left diagram of Figure 6.4, and we assume that lengths $b = AC$, $c = AB$, and $m = AM$ are given, where M is the midpoint of BC. Complete the parallelogram $BACP$ by drawing BP parallel to AC and CP parallel to AB, and draw diagonal AP. Since we know that the diagonals of a parallelogram bisect each other, it follows that AP goes through the midpoint M of BC and that $AP = 2AM = 2m$. Also, we see that $PB = AC = b$. In particular, the lengths of the sides of $\triangle ABP$ are b, c, and $2m$, and therefore $b + c > 2m$.

Now to begin the actual construction, draw a line, and using a compass with memory, mark off segments XN and NQ on this line, each of them having length m. (See the right diagram of Figure 6.4.) Again using a compass with memory, draw a circle centered at X and having radius $c = AB$ and draw a circle centered at Q and having radius $b = AC$. Since $b + c > 2m$, these two circles must intersect, and we select one of the points of intersection and call it Y, as in the figure. We now have $\triangle ABP \cong \triangle XYQ$ by SSS.

Next, complete parallelogram $YXZQ$ by constructing XZ parallel to YQ and QZ parallel to YX. (Recall that we know how to do this by Problem 6.1.) We have now constructed $\triangle XYZ$, and we claim that $\triangle XYZ \cong \triangle ABC$, as desired.

We have $XZ = YQ = b = AC$ and $XY = c = AB$. Also, $\angle BAC = \angle BAP + \angle PAC = \angle BAP + \angle APB$, and similarly, $\angle YXZ = \angle YXQ + \angle XQY$. But $\angle BAP = \angle YXQ$ and $\angle APB = \angle XQY$ since we know that $\triangle ABP \cong \triangle XYQ$. It follows that $\angle BAC = \angle YXZ$, and thus $\triangle ABC \cong \triangle XYZ$ by SAS.   ∎

Next, we set ourselves a harder task.

(6.7) PROBLEM.   Reconstruct a triangle given the lengths of its three medians.

To do this construction, we need to be able to construct the point two thirds of the way from X to Y along a given line segment XY. In fact, a more general construction is available, and we digress to present it before we return to the solution of Problem 6.7.

(6.8) PROBLEM.   Divide a given line segment into n equal parts, where n is any positive integer.

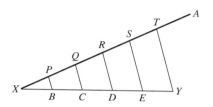

Figure 6.5

Solution.   Let XY be the given segment and draw some line XA different from XY, as shown in Figure 6.5. Select some point P on XA, and using a compass, mark off $n-1$ additional segments PQ, QR, RS, etc. along XA, all of them equal to XP. (We have illustrated the case $n=5$ in Figure 6.5.) Now join the far end of the nth equal segment to Y, as shown in the diagram, and draw PB, QC, RD, etc. parallel to this line. An easy argument with similar triangles now shows that the n pieces XB, BC, CD, etc. into which the original line segment XY is divided all have equal lengths.   ■

Solution to Problem 6.7.   In the original $\triangle ABC$, suppose that the three medians are AM, BN, and CP, and the point where they meet, the centroid of $\triangle ABC$, is G. Now consider $\triangle GBC$. Sides GB and GC of this triangle have lengths $\frac{2}{3}BN$ and $\frac{2}{3}CP$, respectively, and since we are given line segments with lengths BN and CP, we can use Problem 6.8 to construct line segments with lengths GB and GC. Now segment GM is a median of $\triangle GBC$, and since $GM = \frac{1}{3}AM$ and we are given a segment of length AM, we can construct a segment whose length is equal to that of median GM of $\triangle GBC$. We have now constructed the lengths of sides GB and GC and of median GM of $\triangle GBC$, and thus by Problem 6.6, we can reconstruct a triangle congruent to $\triangle GBC$, and in particular, we can construct the length of BC, which is a side of the original triangle. We can similarly construct a line segment whose length is BA, and since we are given a segment whose length is that of median BQ, it follows from Problem 6.6 that we can reconstruct the original triangle.   ■

We will do one more fairly difficult reconstruction problem.

(6.9) PROBLEM.   Reconstruct a triangle given the lengths of its three altitudes.

Solution.   Let h_A, h_B, and h_C be the given lengths of the altitudes from A, B, and C, respectively, in $\triangle ABC$, and as usual, let a, b, and c be the lengths of sides BC, AC, and AB of this triangle. Note that by computing the area of $\triangle ABC$ in three ways, we get the equations $ah_A = bh_B = ch_C$, and thus $h_C/h_A = a/c$ and $h_B/h_A = a/b$. Our strategy will be first to construct a triangle similar to $\triangle ABC$ and then to rescale it so as to obtain a triangle that is actually congruent to $\triangle ABC$.

First, draw line segments PX and PZ of lengths h_B and h_A, respectively, forming $\angle P$, as shown in the left diagram of Figure 6.6. Choose point Y on PX (extended) in some convenient but arbitrary position, as shown, and draw $YW \| XZ$.

Next, we do a similar construction, as shown in the middle diagram of Figure 6.6. This time, we let $QR = h_C$ and $QT = h_A$, and we choose S on QR (extended) so that $RS = XY$. We then draw $SU \| RT$.

The next step is to draw line segment JH so that $JH = XY = RS$ and then to choose point K so that $JK = ZW$ and $HK = TU$. This, of course, is done by taking K to be one of the points of intersection of the circle centered at J and having radius ZW and the circle centered at H and having radius TU. (To be sure that these circles really do intersect, we need to establish that $ZW + TU > JH$. We will explain later why this inequality is guaranteed to hold.)

What have we accomplished? We see that

$$\frac{JH}{JK} = \frac{XY}{ZW} = \frac{PX}{PZ} = \frac{h_B}{h_A} = \frac{a}{b}$$

and

$$\frac{JH}{HK} = \frac{RS}{TU} = \frac{QR}{QT} = \frac{h_C}{h_A} = \frac{a}{c}.$$

It follows that

$$\frac{JH}{a} = \frac{JK}{b} = \frac{HK}{c},$$

and thus $\triangle KHJ \sim \triangle ABC$ by SSS. This also explains why $ZW + TU > JH$: These lengths are proportional to the lengths of the sides of the original triangle.

We have now constructed a triangle similar to our original triangle. As shown in the right diagram of Figure 6.6, construct JD perpendicular to JH, with $JD = h_A$, and then construct DE parallel to JH. Let L be the point of intersection of line JK with line DE and construct LM parallel to KH, where M lies on JH. It follows that $\triangle LMJ \sim \triangle KHJ \sim \triangle ABC$. Furthermore, we have arranged matters so that the altitude from L of $\triangle LMJ$ has length equal to $JD = h_A$, which is the length of

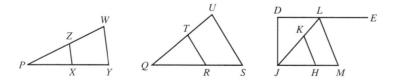

Figure 6.6

the altitude from A in $\triangle ABC$. It is easy to see from this that $\triangle LMJ$ and $\triangle ABC$ are not just similar; they are in fact congruent, as required. ∎

Exercises 6B

6B.1 Given two sides and the altitude to the third side of an acute angled triangle, reconstruct the triangle.

6B.2 Given one side, the altitude to that side, and the median to that side of a triangle, reconstruct the triangle.

6B.3 Given the hypotenuse and the altitude to the hypotenuse of a right triangle, reconstruct the triangle.

6B.4 Given the circumradius, one side, and the altitude to that side of a triangle, reconstruct the triangle.

6C   Tangents

We mentioned earlier that it was not legal to use a straightedge to construct a tangent to a circle simply by placing the straightedge so that it just touches the circle. How, then, can we construct tangents to a circle?

(6.10) PROBLEM.   Given a point on a circle, construct the tangent to the circle at that point.

To solve Problem 6.10, we need to construct the center of a given circle, and so we digress briefly to discuss this easy problem, which appeared as Exercise 6A.3.

(6.11) PROBLEM.   Given a circle, construct its center.

Solution.   Choose three points A, B, and C on the circle. Draw chords AB and AC and construct the perpendicular bisectors of these two chords. The point where the two perpendicular bisectors meet is equidistant from A, B, and C, and hence it is the center of the circle. ∎

Solution to Problem 6.10.   Let P be the given point on the circle. Construct the center O of the circle and draw radius OP. Now erect the perpendicular to OP at P and note that that perpendicular is the desired tangent. ∎

Somewhat more interesting is the following problem.

(6.12) PROBLEM.   Given a circle and a point outside of the circle, construct the two tangents to the circle from the point.

Solution.   Let P be the given point and construct the center O of the given circle. Draw line segment OP and construct its midpoint M. (Note that M may be outside of the given circle, as it is in Figure 6.7, or it may be on or inside the circle.) Draw the circle through P and centered at M, and note that this circle also goes through point O and that segment OP is a diameter. Let A and B be the points where this circle meets the original circle. To avoid

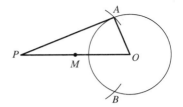

Figure 6.7

clutter in Figure 6.7, we have drawn only small arcs of the circle centered at M, but of course, these arcs are sufficient to define the points A and B. We claim now that line PA, and similarly, also PB, is tangent to the given circle. To see why, observe that it suffices to show that $\angle PAO = 90°$. This is true, however, since $\angle PAO$ is inscribed in the circle centered at M, and line segment PO is a diameter of this circle.   ∎

The construction of a line that is simultaneously tangent to two given circles depends on a subtler trick. (Note that the number of common tangent lines to two given circles can be zero, one, two, three, or four depending on how the two circles are arranged.) We will not attempt to be completely general, and in particular, we will assume that the two given circles have unequal radii. Also, we will construct only a common tangent with the property that it does not separate the centers of the two circles.

(6.13) PROBLEM.   Construct a common tangent line to two given circles.

Solution.   Construct the centers U and V of the two circles, as shown in Figure 6.8, and choose some convenient point Q on the circle centered at V. Construct the "corresponding" point P on the circle centered at U by drawing radius UP parallel to VQ, where P and Q lie on the same side of the line UV joining the centers of the two circles. Let X be the point where PQ meets UV.

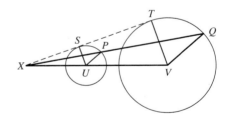

Figure 6.8

We will show that the point X is independent of the choice of Q. To see why this is so, observe that $\triangle XUP \sim \triangle XVQ$, and thus $XU/XV = UP/VQ$. Since UP and VQ are radii of the two given circles, the ratio UP/VQ is clearly independent of the choice of the point Q, and it follows that the ratio XU/XV is also independent of the choice of Q. But it is easy to see that the ratio XU/XV is a monotonically decreasing function of the point X, as X moves along line UV from infinitely far away on the left in Figure 6.8 toward U. It follows that there cannot be two different points X and Y on line UV, both to the left of U and such that

$XU/XV = YU/YV$. This shows that the point X really is independent of Q, as claimed.

We observe next that the common tangent line ST, which is the line we are trying to construct, crosses UV at X. This is because radii US and VT are both perpendicular to ST, and hence these two radii are parallel. In other words, if we took the point Q to be T, then the point P would be S. But by the reasoning in the previous paragraph, we know that for every choice of Q, the line PQ goes through X, and thus in particular, ST goes through X.

We have constructed the point X where the common tangent line ST crosses the line of centers UV. It follows that if we use Problem 6.12 to construct the tangent XS from the point X to the circle centered at U, then that line also must go through T, and thus it is the desired common tangent line. ∎

A more difficult and very general type of problem concerning tangents was studied by Apollonius of Perga about 2200 years ago. These Apollonian problems have the following form: Given three objects, which can be circles, lines, or points, construct a circle tangent to all three objects. What we mean by a circle "tangent to a point" in this context is that the circle should go through the given point. If the given objects are three points, we see that the corresponding Apollonian problem is essentially that of constructing the circumcircle of a given triangle. If the given objects are three lines, on the other hand, the corresponding Apollonian problem is the construction of the inscribed circle of a given triangle. Of course, we have already solved these two easy problems. In the remainder of this section, we discuss several more difficult Apollonian problems. But we shall not present a solution to the most difficult such problem: the construction of a circle tangent to three given circles.

(6.14) PROBLEM. Given two lines and a point, construct a circle through the given point and tangent to the two given lines.

Solution. We give up a little generality and assume that the two given lines are not parallel, and we let R be their point of intersection. In Figure 6.9, the two given lines are RA and RB, and the given point is P. Note that there are actually two circles that go through P and are tangent to lines RA and RB, but we have drawn only one

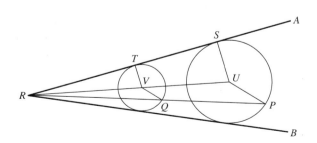

Figure 6.9

of them in Figure 6.9. Our immediate goal will be to construct the center U of the desired circle.

The point U will be equidistant from lines RA and RB, and it follows that U must lie on the bisector of $\angle ARB$. We construct this angle bisector and choose any convenient point V on it. We then drop the perpendicular VT from V to RA, and we draw the circle centered at V and passing through the foot T of this perpendicular. Since RV is the bisector of $\angle ARB$, we see that this circle is tangent to both RA and RB. But of course, this circle probably does not pass through the point P; we still have more work to do.

Next, we choose one of the two points where line RP meets the circle centered at V, and we call this point Q. (This choice determines which of the two circles through P that are tangent to RA and RB we eventually obtain.) We construct the line through P parallel to QV, and we let U be the point where this line meets the angle bisector RV, as shown in the diagram.

Since by construction, U lies on the bisector of $\angle ARB$, we know that the circle centered at U and tangent to RA will also be tangent to RB. We construct this circle by dropping the perpendicular US from U to RA and drawing the circle through S and centered at U. To prove that this circle also goes through point P, as required, we must show that $UP = US$.

Now $VT \| US$ since both VT and US are perpendicular to AR, and also $VQ \| UP$ by construction. It follows that $\triangle TVR \sim \triangle SUR$ and also that $\triangle QVR \sim \triangle PUR$. We thus have

$$\frac{VQ}{UP} = \frac{VR}{UR} = \frac{VT}{US},$$

and thus $UP/US = VQ/VT$. But $VQ = VT$ since Q and T lie on the same circle centered at V, and hence $UP = US$, as required.   ∎

Now that we have solved Problem 6.14, a related Apollonian problem becomes fairly easy.

(6.15) PROBLEM.   Given two lines and a circle, construct a circle tangent to the three given objects.

Solution.   The two given lines appear in Figure 6.10 as RA and RB, which we are assuming are not parallel. The given circle is the one centered at P in the diagram, and we will say that the radius of this circle is r. Construct line SX parallel to RA, as shown, where the distance between RA and SX is equal to r. To carry out the construction of SX, several easy steps are needed. First, erect a perpendicular to RA at some point (say, at R), and then using a compass with memory, mark off on this perpendicular line a distance from R equal to r. Finally, draw SX parallel to RA through the marked point. To do this, one must, of course, first find the distance r by constructing the point P, which is the center of the given circle. Similarly, construct SY as shown, where SY is parallel to RB and the distance between these parallel lines is also equal to r.

Now, using the construction of Problem 6.14, find the center U of a circle tangent to SX and SY and running through point P. Then the distances from U to SX, SY, and P are all equal, and the distance from U to the given circle is $UP - r$. We see that $UP - r$ is also the distance from U to each of the lines RA and RB, and it follows that U is the center of the circle that we seek. A point lying on this circle is the intersection of line segment UP with the given circle, and so we can complete our construction by drawing the circle through that point and centered at U. ∎

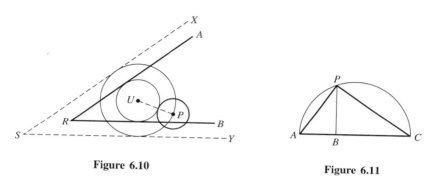

Figure 6.10   **Figure 6.11**

Before we present our final Apollonian problem, we digress briefly to develop a tool that we will need. Given three quantities x, y, and z, we say that y is the **mean proportional** between x and z if $x/y = y/z$, or equivalently, $y^2 = xz$.

(6.16) PROBLEM.   Construct a length equal to the mean proportional of two given lengths.

Solution.   We can measure off the two given lengths using a compass with memory as adjacent segments of some line, and so in Figure 6.11, we assume that AB and CB are the two given lengths. Draw the circle with diameter AC and erect a perpendicular BP to AC at B, where P lies on the circle. As shown in the diagram, we really need only the semicircle.

We claim that BP is the mean proportional between AB and CB. To see why this is true, observe that $\angle APC = 90° = \angle CBP$, and thus $\angle PAB = 90° - \angle PCA = \angle CPB$. Since also $\angle ABP = 90° = \angle PBC$, we see that $\triangle ABP \sim \triangle PBC$ by AA, and hence $AB/PB = PB/CB$. Thus PB is the mean proportional between AB and CB, as desired. ∎

We can now present and solve our final Apollonian problem.

(6.17) PROBLEM.   Construct a circle through two given points and tangent to a given line.

Solution.   We assume that the line determined by the given points P and Q is not parallel to the given line AB, and we construct the intersection point U of line PQ

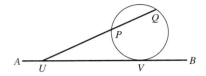

Figure 6.12

with line AB, as shown in Figure 6.12. Our immediate goal is to construct the point of tangency V of the desired circle with line AB. Once V is found, we can construct the circle as the circumcircle of $\triangle PQV$. To find V, of course, it suffices to construct the distance UV.

Recall from Theorem 3.21 that we have $UP \cdot UQ = (UV)^2$. (See also Theorem 1.35.) Thus $UP/UV = UV/UQ$, and so the required distance UV is the mean proportional between UP and UQ. We can construct this distance by Problem 6.17.   ∎

Exercises 6C

6C.1 Given a circle and a line, construct a line tangent to the given circle and parallel to the given line.

6C.2 Given two circles whose interiors have no points in common, construct a line tangent to both circles and such that the centers of the circles lie on opposite sides of the line.

6C.3 Given a square and a positive integer n, construct a square whose area is exactly n times the area of the given square.

6C.4 Given two lines and a point on one of them but not on the other, construct a circle through the point that is tangent to both lines. Consider both the case where the given lines are parallel and the case where they intersect.

6C.5 Construct a circle tangent to two given circles and having a radius equal to the length of a given line segment. Assume that the given line segment is long enough to make this possible.

6D   Three Hard Problems

The ancient Greek geometers proposed three construction problems that became notorious for their difficulty: squaring a circle, doubling a cube, and trisecting an angle. In fact, these problems remained unresolved until comparatively modern times, but it is now known that each of these three constructions is impossible. In this section, we will discuss these three classical hard construction problems, and in the following section, we will try to indicate without giving detailed formal proofs why they are impossible.

What does it mean to square a circle? In general, to **square** a geometric figure means to construct a square whose area is equal to the area of the given figure.

(6.18) PROBLEM. Square a given triangle.

Solution. Designate one side of the given triangle as the base and construct its midpoint. If b is the length of the base, therefore, we have constructed a line segment whose length is $b/2$. Now construct the altitude of the given triangle perpendicular to the designated base and let h be its height. We know that the area of the triangle is $\frac{1}{2}bh$, and thus our task is to construct a square having this same area. If s is the length of the side of the desired square, therefore, we have $s^2 = \frac{1}{2}bh$, and thus s is the mean proportional between $b/2$ and h. We have line segments with lengths $b/2$ and h, and by Problem 6.16, we know how to construct a line segment whose length is the mean proportional of these two lengths. We can therefore construct a line segment of length s, and once we have that, it is easy to construct a square having that line segment as a side.
∎

What figures other than triangles can be squared? If we wish to square a quadrilateral, for example, we could draw a diagonal to divide it into two triangles, and then using Problem 6.18, we could construct two squares, each having the same area as one of the two component triangles of the given quadrilateral. It suffices, therefore, to be able to construct a square whose area is equal to the sum of the areas of two given squares.

(6.19) PROBLEM. Given two squares, construct a square whose area is the sum of the areas of the two given squares.

Solution. This is very easy. Simply construct a right triangle whose arms have lengths equal to the sides of the two given squares and then construct a square whose side is the hypotenuse of this right triangle. By the Pythagorean theorem, the area of the square on the hypotenuse is the sum of the areas of the two given squares.
∎

Similar reasoning allows us to square any polygon. Suppose, for example, we have a polygon with n sides, where $n > 3$ is an integer. If we work by mathematical induction, we can suppose that we already know how to square polygons with fewer than n sides. To square the given n-gon, draw a diagonal that divides the figure into two polygons, each having fewer than n sides, and square each of these. Then use Problem 6.19 to construct a square whose area is the sum of the areas of the two squares just constructed.

Actually, we are cheating a little here. How do we know that it is always possible to find a diagonal of a polygon that divides it into two smaller polygons? If the original polygon is convex, then it is easy to see that any of its $n(n-3)/2$ diagonals will work. (Recall that a polygon is **convex** if none of its angles exceeds 180°, and in this situation, all of the diagonals lie inside the figure.) But if the polygon is not convex, then at least some of its diagonals do not lie inside the figure. Nevertheless, even a nonconvex polygon always has at least one interior diagonal, and so that diagonal can be used to subdivide the polygon into two smaller polygons. The fact that such an interior diagonal must always exist is not obvious, however, especially if the polygon has a very large number of sides.

Since circles are surely the simplest nonpolygonal figures, and since we now know how to square any polygon, it is reasonable to try to find a technique that will square a circle. But in fact, it is not possible to do this. It is a theorem that no construction using only a straightedge and compass, and obeying the official rules, can succeed in squaring a circle.

We stress that we are saying much more than that no one knows how to square a circle: It is definitely known that squaring a circle is impossible. Unfortunately, this distinction is sometimes misunderstood, and there are people who continue to attempt to solve this and the other two impossible Greek construction problems. There are, in fact, some individuals who insist that they have succeeded in solving one or more of these problems and who refuse to accept the fact that their solutions cannot be correct. University mathematics departments occasionally receive letters from self-proclaimed circle squarers and cube doublers, and most often, from angle trisectors. These people usually seek recognition, and sometimes even money, for their "accomplishment." Usually, the proposed, and necessarily incorrect, construction fails either because it does not follow the official rules or because even if it could be carried out perfectly, using an ideal straightedge and an ideal compass, it would result in only an approximate solution to the problem. (Of course, any actual construction, carried out with real physical tools, can at best yield only an approximate solution, but the correct constructions that we have been discussing, if they could be carried out with perfect, ideal tools, would always yield exact solutions.)

In Problem 6.19, we saw how to use the Pythagorean theorem to construct a square whose area is equal to the sum of the areas of two given squares. In particular, we can construct a square whose area is double that of a given square. Probably, the easiest way to do this is to construct the square whose side is a diagonal of the original square. We cannot resist presenting a very slick proof that this works. This proof does not rely on an appeal to the Pythagorean theorem.

In Figure 6.13, the given square is $ABCD$ and we have drawn square $AEFC$, whose side is diagonal AC of the original square. It is obvious in the figure that the small square is composed of two congruent copies of $\triangle ABC$, while the large square is composed of four copies of this triangle. It follows that the large square has exactly double the area of the small square, as claimed. Of course, it is necessary to prove that the five small triangles in Figure 6.13 actually are congruent, but that is easy, and we omit the details.

We now know that it is easy to construct a square whose area is exactly double the area of a given square. Analogously, the Greek geometers posed the problem of constructing a cube whose volume is double the volume of a given cube. Of course, since a cube is a three-dimensional object, we cannot actually be given a cube on our piece of paper. What we are really given is a line segment whose length equals the edge length of the given cube, and the task is to construct a line segment equal to the edge of a cube with double the volume. If the length of our given line segment is s, then the volume of the original cube is s^3, and double that is $2s^3$. The edge length of the doubled cube is, therefore, $\sqrt[3]{2s^3} = \sqrt[3]{2}s$. In other words, the problem of doubling a cube

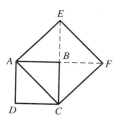

Figure 6.13

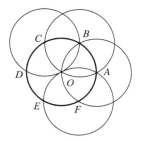

Figure 6.14

requires that we construct with straightedge and compass a line segment whose length is $\sqrt[3]{2}$ times the length of a given segment. As was the case with squaring a circle, this is provably impossible. It is a theorem that no legal construction can accomplish this feat.

The third classical hard construction problem is that of trisecting a given angle. As we have seen, it is easy to bisect a given angle, and by Problem 6.8, we know that we can trisect a given line segment, and so it certainly seems reasonable that one should be able to trisect a given angle. Some angles, in fact, can be trisected. To trisect a 90° angle, for example, it suffices to construct a 30° angle, and this is very easy: Simply construct an equilateral triangle and bisect one of its angles. But it is impossible to trisect an *arbitrary* given angle using only a straightedge and compass.

In fact, an even stronger impossibility theorem is true. Not only does there not exist a straightedge and compass procedure that will trisect an arbitrary angle, but it is actually impossible to trisect a certain particular angle, namely, 60°. If it were possible to trisect a 60° angle, then we could easily construct a 40° angle: Simply construct an equilateral triangle and trisect one of its angles, thereby dividing it into a 20° angle and a 40° angle. As we shall explain, however, it is not possible to construct a 40° angle, and thus it is not possible to trisect a 60° angle. It is, therefore, certainly impossible to trisect an arbitrary angle.

We know how to construct a 90° angle and a 60° angle, and we shall see how to construct a 36° angle, but it is impossible to construct a 40° angle. What is going on here? A good way to think about this is to consider the following general problem: Divide a circle into n equal arcs, where n is some positive integer. Of course, we require that this division of a circle should be carried out using only a straightedge and compass and by following the official construction rules.

The circle division problem is trivial if $n = 1$ and it is very easy if $n = 2$: Construct the center of the circle and draw a diameter. The two points where the diameter meets the circle divide the circle into two 180° arcs. The case $n = 4$ is also easy: Draw two perpendicular diameters to obtain four 90° arcs. Once the center O of the circle has been constructed, it is also easy to divide the circle into six equal arcs. In fact, this can be done with a compass without memory and without using a straightedge.

As shown in Figure 6.14, we mark an arbitrary point A on the circle and then draw the circle through O centered at A and meeting the original circle at points B and F. Next, we draw the circle through O and centered at B, and we note that since $AB = OA = OB$, this circle goes through A and also meets the original circle at C.

Similarly, we draw the circles through O centered at F and B to obtain two new intersection points: E and C. Finally, we construct the point D by drawing the circle through O and centered at C.

Since $\triangle AOB$ is equilateral, we see that $\overset{\circ}{\overarc{AB}} = \angle AOB = 60°$, and similarly, each of arcs $\overarc{BC}$, $\overarc{CD}$, $\overarc{EF}$, and $\overarc{FA}$ measures $60°$. Since the whole circle consists of $360°$ of arc, it follows that $\overarc{DE}$ must also be $60°$, and thus the points A, B, C, D, E, and F divide the circle into six equal arcs.

Now that we have solved the circle division problem for $n = 6$, we see that we get for free a solution for $n = 3$: Just take points A, C, and E in Figure 6.15. More generally, we see that if we can solve the circle division problem for any even integer $n = 2m$, then we can also solve it for $n = m$ by taking alternate division points. Conversely, if we can solve the circle division problem for some integer m, then we can also solve it for $n = 2m$. To do this, it suffices to construct the midpoint of a given arc, and this is easy, since we can bisect the corresponding central angle. In particular, we now see that the circle division problem can be solved for all integers of the form $n = 2^e$ and all integers of the form $n = 3 \cdot 2^e$, where $e \geq 0$.

If a circle is divided into n equal arcs, then of course, each of these arcs will measure $360°/n$, and so each of the corresponding central angles will also measure $360°/n$. It follows that we can divide a circle into n equal arcs if and only if we can construct an angle equal to $360°/n$. Since $360/9 = 40$, we see that the impossibility of constructing a $40°$ angle is equivalent to the assertion that the circle division problem cannot be solved when $n = 9$.

We mention that if we divide a circle into n equal arcs, with $n \geq 3$, and we join adjacent division points with line segments, we obtain a regular n-gon. (Recall that a polygon is **regular** if all of its sides are equal and all of its angles are equal.)

We have seen that the circle division problem can be solved when n is one of the numbers 1, 2, 3, 4, 6, 8, 12, 16, 24, etc., and we have said that it cannot be solved when $n = 9$. The obvious question, of course, is: What exactly is the full set of positive integers n for which the problem can be solved? The complete answer was found about 200 years ago by Carl Gauss, who was one of the greatest mathematicians who ever lived.

To describe the set of integers for which the circle division problem can be solved, we need to digress briefly into a discussion of prime numbers. This seems appropriate in this book on Euclidean geometry since Euclid, himself, made one of the earliest significant contributions to the theory of prime numbers: He proved that there are infinitely many of them.

Recall that an integer $p > 1$ is said to be **prime** if its only divisors are 1 and itself. And the number 1 is, by definition, not prime. The first several prime numbers are 2, 3, 5, 7, 11, 13, and 17, and it should be clear that with the exception of the prime number 2, all prime numbers are odd. The mathematician P. Fermat considered the question of which odd prime numbers can be written in the form $p = 1 + 2^e$, where e is a positive integer, and in his honor, such primes are called **Fermat primes**. If we search for Fermat primes by computing the numbers of the form $1 + 2^e$ for the first dozen or so positive integers e, we find the following Fermat primes: $3 = 1 + 2^1$, $5 = 1 + 2^2$, $17 = 1 + 2^4$, and $257 = 1 + 2^8$. The number $1 + 2^e$ is prime when the exponent e is one of the numbers 1,

2, 4, or 8, and the apparent pattern that we see here suggests that perhaps we might also get a prime when $e = 16$. In fact, we do; the number $1 + 2^{16} = 65537$ is indeed prime.

It is not hard to prove that the only way that the number $1 + 2^e$ can possibly be prime is when e is a power of 2, and thus all Fermat primes must have the form $1 + 2^{2^a}$ for integers $a \geq 0$. The numbers $F_a = 1 + 2^{2^a}$ are called **Fermat numbers**, and although it is true that every Fermat prime is a Fermat number, it is certainly not true that every Fermat number is prime. We have $F_0 = 3$, $F_1 = 5$, $F_2 = 17$, $F_3 = 257$, and $F_4 = 65537$, and of course, Fermat knew that these five numbers are prime. Fermat was unable to factor $F_5 = 4294967297$, but it was later found that this number is not prime; it is a multiple of 641. It is now known that none of the next several Fermat numbers is prime, and in fact, no one has found any Fermat primes other than the five that were known to Fermat. The only known Fermat primes, therefore, are 3, 5, 17, 257, and 65537, but it is not known whether or not these are all of the Fermat primes.

We can now state Gauss' theorem, but unfortunately we will not be able to give the proof in this book.

(6.20) THEOREM.   *The circle division problem is solvable for an integer n if and only if $n = 2^e \cdot m$, where $e \geq 0$ is an integer and m is either equal to 1 or else m is a product of different Fermat primes.*

In other words, to decide whether or not the circle division problem is solvable for some integer n, first factor out from n as many factors of 2 as possible and write $n = 2^e \cdot m$, where e is a nonnegative integer and m is an odd number. Next, factor m into prime numbers and check that all of the prime factors of n are distinct and that all of them are Fermat primes.

Since there are only five known Fermat primes, we see that there are exactly 32 known numbers that can play the role of m in Gauss' theorem. The first several of these are 1, 3, 5, $15 = 3 \cdot 5$, 17, $51 = 3 \cdot 17$, $85 = 5 \cdot 17$, $255 = 3 \cdot 5 \cdot 17$, and 257. In particular, because Gauss' theorem does not allow the possibility that $m = 9$, it follows that the circle division problem cannot be solved for $n = 9$, and thus a 40° angle cannot be constructed and hence a 60° angle cannot be trisected. Thus Gauss' theorem implies that it is impossible to trisect an arbitrary given angle with straightedge and compass.

Because Gauss' theorem allows the possibility that $m = 5$, it follows that the circle division problem can be solved for $n = 5$, and this fact was known to the ancients. It was not known, however, that the problem could be solved when $n = 17$, and hence that it was possible to construct a regular 17-gon with straightedge and compass. Gauss was the first person to accomplish that feat.

We close this section by showing how to solve the circle division problem with $n = 10$. By taking alternate points, this also solves the circle division problem for $n = 5$, and if we join these five division points, we will have constructed a regular pentagon.

(6.21) PROBLEM.   Divide a circle into ten equal arcs and construct a regular pentagon.

Solution.   We want to construct an arc of 36° in our circle, which we can assume has radius 1. Our first task will be to determine the length x of the chord that will cut off an arc measuring 36°, and for this purpose, we consider two radii AB and AC that form a 36° central angle $\angle BAC$. Then $\triangle ABC$ is isosceles, where radii $AB = 1 = AC$ and $\angle BAC = 36°$, and the base BC is the chord whose length x we are trying to compute. Note that each of the base angles of $\triangle ABC$ is $(180° - 36°)/2 = 72°$.

Let BD be the bisector of $\angle ABC$, as shown in Figure 6.15, and observe that $\angle ABD = 36°$. Thus $\triangle ABD$ is isosceles, and we have $BD = AD$. Also, since $\angle DBC = 36°$ and $\angle C = 72°$, we see that $\angle BDC = 72°$, and hence $\triangle BDC$ is isosceles and $BD = BC = x$. It follows that $AD = x$, and thus $DC = 1 - x$, as indicated in the figure.

But $\triangle ABC \sim \triangle BDC$ by AA, and it follows that

$$\frac{x}{1} = \frac{BD}{AB} = \frac{DC}{BC} = \frac{1 - x}{x},$$

and thus $x^2 = 1 - x$. We have $x^2 + x - 1 = 0$, and when we solve this equation using the quadratic formula, we obtain $x = (-1 \pm \sqrt{5})/2$. Of course, the length x cannot be negative, and hence $x = (-1 + \sqrt{5})/2$.

To carry out the desired construction now, it suffices to construct a line segment of this length. We start with a circle and assume that the radius is 1 unit. Construct the center A, choose a point B on the circle, draw radius AB, construct a perpendicular radius AP, construct the midpoint M of AP, and draw MB. All of this is shown in Figure 6.16.

Now $AB = 1$ and $AM = 1/2$, and thus $BM = \sqrt{5}/2$ by the Pythagorean theorem. Now swing an arc through A centered at M and let Q be the point where this arc crosses segment MB, as shown in the figure. Then $BQ = BM - QM = BM - AM = \sqrt{5}/2 - 1/2$, and so the length of BQ is the number x that we calculated previously. Next, swing an arc through Q and centered at B and let C be a point where this arc cuts the circle. Then chord BC has length $BQ = x$, and so this chord subtends a central angle of 36°, and we have $\overset{\circ}{\overarc{BC}} = 36°$. Next, we construct the point D, where the circle through B and centered at C meets the

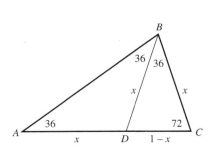

Figure 6.15

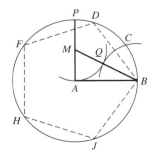

Figure 6.16

original circle. Then $\overset{\circ}{\overparen{CD}} = 36°$, and we continue like this, moving around the circle in jumps of 36° and thereby dividing the circle into ten equal arcs. If we join alternate division points with line segments, as shown in the figure, we get a regular pentagon.

∎

Exercises 6D

6D.1 Divide a circle into 15 equal arcs.

6D.2 Construct an equilateral triangle whose area is equal to that of a given square.

6D.3 Suppose that e is a positive integer and that $p = 1 + 2^e$ is a prime number. Prove that e must be a power of 2.

> *HINT:* If e is not a power of 2, it is possible to factor $e = ab$, where a is odd and $1 < a$. Let $m = 1 + 2^b$ and note that $p = (m - 1)^a + 1$ is a multiple of m. This yields a contradiction.

6D.4 Compute the length of a diagonal of a regular pentagon whose side has length 1.

6E   Constructible Numbers

In Section D, we asserted without proof that it is impossible to find straightedge and compass constructions that will square a circle, double a cube, or trisect an angle. Although we argued that the impossibility of the trisection problem is a consequence of the fact that one cannot construct a 40° angle, we did not really explain why the construction of such an angle is impossible. (That a 40° angle cannot be constructed is a consequence of the much more general theorem of Gauss on circle division, but we certainly did not explain why Gauss' theorem is true.) We will now try to indicate some of the ideas that underlie the impossibility proofs for the three classical impossible problems.

Recall that if we could double a cube, then we could solve the following problem: Given an arbitrary line segment, construct a segment whose length is $\sqrt[3]{2}$ times the length of the given segment. In fact, this is precisely what the cube doubling problem requires us to do. With this in mind, we define what we shall refer to as the **ratio construction problem** for a given positive number α: Given any line segment, construct a segment whose length is α times the length of the given segment. For example, the ratio construction problem is completely trivial when $\alpha = 1$, and it is very easy when α is a positive integer. Also, since it is possible to double a square, it follows that the ratio construction problem is solvable when $\alpha = \sqrt{2}$, and by Problem 6.8, we know that the ratio construction problem is also solvable when $\alpha = 1/n$, where n is any positive integer.

Suppose now that there exists a construction technique that will square a circle. Given any line segment, we can draw a circle having that segment as a radius, and then we can square that circle to obtain a square with side s. (Recall that we are assuming it is possible to square a circle.) If the length of the given line segment is r, then the area of the circle is πr^2, and since the square and the circle have equal areas, we have $s^2 = \pi r^2$. Thus $s = \sqrt{\pi} r$, and we have constructed a line segment whose length is $\sqrt{\pi}$ times the length of the given segment. In other words, if we could square a circle, we could solve the ratio construction problem with $\alpha = \sqrt{\pi}$.

Finally, suppose that it is possible to construct a 40° angle. Given any line segment, we can construct a right $\triangle ABC$, where the hypotenuse BC is the given segment and where $\angle B = 40°$. Then $AB/BC = \cos(40°)$, and thus the length of segment AB is $\cos(40°)$ times the length of the given segment. This shows that if it were possible to trisect angles, and thereby to construct a 40° angle, then we could solve the ratio construction problem for $\alpha = \cos(40°)$.

We see now that to prove that the three constructions we have been discussing are impossible, it suffices to show that the ratio construction problem cannot be solved when α is any of the numbers $\sqrt[3]{2}$, $\sqrt{\pi}$, or $\cos(40°)$. We want, therefore, to try to understand exactly for which positive numbers the ratio construction problem is solvable. We say that such numbers are **constructible**, and our goal is to show that none of the three numbers $\sqrt[3]{2}$, $\sqrt{\pi}$, and $\cos(40°)$ is constructible.

For technical reasons, it is convenient to expand our definition of constructible real numbers to include 0 and to include those negative numbers whose absolute values are constructible. With this understanding, it is easy to show that the set C consisting of all constructible numbers is closed under addition and subtraction and multiplication. It is only slightly harder to show that the constructible number set C is also closed under division, but of course, we must not divide by zero. In other words, C is what is called a **subfield** of the field of real numbers. This explains why the branch of abstract algebra called field theory is the basic tool for establishing that a particular number is or is not constructible.

Although we shall omit the proofs of most of the facts that we assert about C, we remind the reader that these are indeed theorems, and to emphasize this point, we will prove a few of them. Also, although we will not go deeply into field theory, we will try to give the reader at least the flavor of what is involved.

We begin by proving one of the facts that we mentioned previously.

(6.22) THEOREM.   *The set C of constructible numbers is closed under division.*

Proof.   Suppose that α and β lie in C and assume, as we may, that α and β are positive. We are given a line segment AB, and our goal is to construct a line segment whose length is $\frac{\alpha}{\beta} AB$.

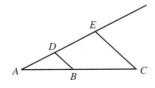

Figure 6.17

As in Figure 6.17, we draw a line through A making some convenient angle with the given line AB. Since β is constructible, we can construct a line segment of length $\beta \cdot AB$, and so using a compass with memory, we can construct the point D, as in the figure, with $AD = \beta \cdot AB$. Also, since α is constructible, we can construct the point E on AD, as shown, with $AE = \alpha \cdot AB$.

Next, draw DB and then construct $EC \parallel DB$, where C lies on line AB. It follows that $AE/AC = AD/AB$, and thus

$$AC = \frac{AE \cdot AB}{AD} = \frac{\alpha \cdot (AB)^2}{\beta \cdot AB} = \frac{\alpha}{\beta} AB \,,$$

and hence line segment AC has the desired length. ∎

Not only is C closed under the ordinary arithmetic operations of addition, subtraction, multiplication, and division, but it is also closed under the extraction of real square roots.

(6.23) THEOREM. *If α is in C and $\alpha > 0$, then $\sqrt{\alpha}$ also lies in C.*

Proof. We are given a line segment of length x, and we know that we can construct a segment of length αx. Since $\sqrt{\alpha} x$ is the mean proportional between x and αx, we can use Problem 6.16 to construct a line segment of length $\sqrt{\alpha} x$, and this shows that $\sqrt{\alpha}$ is constructible, as required. ∎

We now know that complicated looking numbers such as

$$\frac{\sqrt{\dfrac{3 - \sqrt{5}}{2} - 3}}{2 - \sqrt{2}}$$

lie in C. More precisely, any number that can be formed by starting with integers and repeatedly (but finitely often) adding, subtracting, multiplying, dividing, and extracting square roots lies in C. Of course, we are only allowed to extract square roots of positive numbers, and we are not allowed to divide by zero.

It is a deeper fact that every member of C can be formed in this way: by repeatedly using ordinary arithmetic operations together with the extraction of real square roots. Field theoretically, what we are saying here is that C is the unique smallest subfield of the real numbers that is closed under the extraction of real square roots. It is also true that C is exactly the real subfield of the square-root closure of the rational numbers in the complex numbers, but we shall not need this slightly deeper fact.

Although we will not provide any details of the proof here, it is easy to understand the underlying reason for the assertion of the previous paragraph. At every step in a legal geometric construction, we are either taking the intersection of two lines, the intersection of two circles, or the intersection of a line and a circle. If we were to work out the coordinates of a point obtained in this way, we would have to solve some equations, but these equations can never be worse than quadratic. Essentially, this is because the equation of a circle contains no power higher than the second. But because of the quadratic formula, we know that linear and quadratic equations can be solved by the processes of ordinary arithmetic together with the extraction of square roots. It can be proved from this that every number in C is formed by repeatedly applying these operations.

To show that squaring a circle, doubling a cube, and constructing a $40°$ angle are impossible, we need to establish that $\sqrt{\pi}$, $\sqrt[3]{2}$, and $\cos(40°)$ do not lie in C, and so we need to show that none of these numbers can be formed by arithmetic operations together with square root extraction.

To understand the underlying principle of the proof, we recall that if $f(x)$ is a polynomial, then a **root** of the polynomial f is a number x such that $f(x) = 0$. For example, the numbers $2/3$, $\sqrt[3]{2}$, and $-1 + \sqrt{5}$ are roots of the polynomials $3x - 2$, $x^3 - 2$, and $x^2 - 2x + 4$, respectively. A real or complex number is said to be **algebraic** if it is a root of some nonzero polynomial having integer coefficients, and thus, for example, each of the numbers $2/3$, $\sqrt[3]{2}$, and $-1 + \sqrt{5}$ is algebraic. But not every number is algebraic, and those that are not algebraic are said to be **transcendental**. If one knows the trick, it is not very hard to prove that transcendental numbers actually exist. (The trick, discovered by G. Cantor (1845–1918), is to show that the set of real numbers is uncountably large, while the set of algebraic numbers is merely countably infinite.)

What is much more difficult is to establish that some particular number is not algebraic. It is a famous theorem of C. Lindemann (1852–1939), for example, that π is transcendental, and this result is considered a major accomplishment of 19-century mathematics. (It is also known that e is transcendental, but to demonstrate how hard such problems can be, we mention that it is still unknown whether or not the number $\pi + e$ is algebraic.)

One can show that any number built from algebraic numbers by repeated addition, subtraction, multiplication, division, and extraction of roots is itself algebraic, and so in particular, the members of C are all algebraic. If a circle could be squared, then $\sqrt{\pi}$ would lie in C, and thus π would also lie in C since C is closed under multiplication. But by Lindemann's theorem, π is not algebraic, and hence it cannot lie in C. This contradiction shows that circles cannot be squared.

Recall that the **degree** of a nonzero polynomial $f(x)$ is the highest power of x that appears with a nonzero coefficient in the polynomial. Thus, for example, the polynomial $2x^3 - 4x + 5$ has degree 3. If α is an algebraic number, then by definition, there is some nonzero polynomial $f(x)$ with integer coefficients such that $f(\alpha) = 0$, and we define the **degree** of α to be the smallest possible degree of such a polynomial. The key fact here, proved by elementary field theory, is that since the members of C can all be built using arithmetic and the extraction of square roots, it follows that the degree of every member of C is a power of 2. To prove that some algebraic number α does not lie in C, therefore, it suffices to show that the degree of α is not a power of 2, and so it would be useful to be able to compute this degree. (We should mention, however, that not every real algebraic number whose degree is a power 2 actually lies in C. In other words, the condition that the degree is a power of 2 is necessary but it is not sufficient.)

Suppose that $f(\alpha) = 0$, where f is a nonzero polynomial whose degree equals the degree of the algebraic number α. In other words, we are assuming that the polynomial f has the smallest possible degree such that $f(\alpha) = 0$. We claim that in this case, $f(x)$ cannot be factored as $g(x)h(x)$, where neither $g(x)$ nor $h(x)$ is constant. To see why, observe that if $f(x) = g(x)h(x)$ and neither $g(x)$ nor $h(x)$ is constant, then each of $g(x)$ and $h(x)$ must have degree smaller than the degree of $f(x)$. But by the minimality of the degree of f, we cannot have $g(\alpha) = 0$ or $h(\alpha) = 0$, and this is a contradiction since $0 = f(\alpha) = g(\alpha)h(\alpha)$.

This converse of this fact is also true, but it is a bit harder to prove. If $f(\alpha) = 0$ and the nonzero polynomial f has no factorization into two nonconstant polynomials with integer coefficients, then the degree of f is equal to the degree of α. This provides a tool that we can use to compute the degrees of certain algebraic numbers.

Consider, for example, the number $\alpha = \sqrt[3]{2}$. Clearly, α is a root of the degree 3 polynomial $f(x) = x^3 - 2$, and so our task is to determine whether or not this polynomial can be factored as a product of two nonconstant polynomials with integer coefficients. Assuming that there is such a factorization, we can write $x^3 - 2 = (ax^2 + bx + c)(dx + e)$, where $a, b, c, d,$ and e are integers. It follows that $ad = 1$ and $ce = -2$, and thus $d = \pm 1$ and e has one of the four values ± 1 or ± 2. Since $dx + e$ is a factor of $f(x)$, however, it follows that $f(-e/d) = 0$. But $-e/d$ is one of the numbers ± 1 or ± 2, and it is a triviality to check that none of these four numbers is a root of f. This contradiction shows that our assumed factorization cannot exist, and thus the degree of the algebraic number $\sqrt[3]{2}$ is 3. Since 3 is not a power of 2, we conclude that $\sqrt[3]{2}$ is not constructible, and hence it is impossible to double a cube using only a straightedge and compass.

Finally, we deal with the number $\alpha = \cos(40°)$. We want to know whether or not α is algebraic, and if it is, we want to determine its degree. The key to this is the following triple-angle formula for cosines. With a little manipulation of trigonometric identities, we have

$$
\begin{aligned}
\cos(3\theta) &= \cos(\theta + 2\theta) \\
&= \cos(\theta)\cos(2\theta) - \sin(\theta)\sin(2\theta) \\
&= \cos(\theta)(\cos^2(\theta) - \sin^2(\theta)) - \sin(\theta)(2\sin(\theta)\cos(\theta)) \\
&= \cos^3(\theta) - 3\cos(\theta)\sin^2(\theta) \\
&= 4\cos^3(\theta) - 3\cos(\theta),
\end{aligned}
$$

where we obtained the last equality by substituting $1 - \cos^2(\theta)$ for $\sin^2(\theta)$. Since $\cos(120°) = -1/2$, we can apply this formula with $\theta = 40°$ to obtain $4\alpha^3 - 3\alpha = -1/2$, and thus $8\alpha^3 - 6\alpha + 1 = 0$. It follows that α is a root of the polynomial $f(x) = 8x^3 - 6x + 1$, and in particular, α is algebraic.

Suppose that there exists a factorization $f(x) = (ax^2 + bx + c)(dx + e)$, where $a, b, c, d,$ and e are integers. Then $ad = 8$ and $ec = 1$, and we see that there are only eight possibilities for the number $-e/d$, and these are $\pm 1, \pm 1/2, \pm 1/4,$ and $\pm 1/8$. But since $dx + e$ is a factor of $f(x)$, we know that $-e/d$ must be a root. A routine check shows that none of these eight possibilities for $-e/d$ actually is a root of f, and it follows that the assumed factorization of f cannot exist. By our earlier remarks, we conclude that the degree of $\cos(40°)$ is 3, and thus $\cos(40°)$ is not a constructible number since 3 is not a power of 2. We conclude that it is indeed impossible to construct a 40° angle, and hence it is impossible to trisect an arbitrary given angle using only a straightedge and compass.

Exercises 6E

6E.1 If α and β lie in the set C of constructible numbers prove that $\alpha\beta$ is constructible.

6E.2 Is it possible to construct with straightedge and compass a line segment whose length is equal to the circumference of a given circle? Explain how you know.

6F   Changing the Rules

In this section, we discuss some of the constructions that can be done using nonstandard tools or by using the usual tools and following nonstandard rules. We begin with a pretty angle trisection construction attributed to Archimedes (287?–212 B.C.). Although Archimedes' trisection construction uses only a straightedge and compass, it does not follow the usual official rules because it requires marks to be placed on the straightedge.

Suppose that the angle we want to trisect is $\angle ABC$, as shown in Figure 6.18. Choose a point P on side BC of the angle, draw the circle through P and centered at B, and extend side AB, as shown in the figure.

The next task is to draw the line labeled PR in the figure, where R lies on the line AB and where PR meets the circle at a point Q such that QR is equal to the radius of the circle. Although it is impossible to construct this line if we limit ourselves to the usual tools and rules, it is easy with the following procedure. First, we make two marks on our straightedge, positioned so that the distance between them is equal to the radius of the circle. We then place our marked straightedge on the paper in such a way that it runs through the point P and so that the mark farthest from P is at the point where the straightedge crosses line AB. The dashed line in Figure 6.18 shows a possible position for the straightedge, where the two marks are indicated by the arrowheads. We then slowly turn and slide the straightedge so that it continues to run through the point P, while the mark that started on line AB remains on that line and moves toward the point B. An easy way to picture this turn-and-slide process is to imagine that a pin is stuck into the paper at P. We move the mark along line AB while keeping the straightedge pressed against the pin. We continue this turning and sliding of the straightedge until the second mark just touches the circle. (Note that if the straightedge starts in the position indicated by the dashed line in the figure, we want to turn it counterclockwise and slide it upward

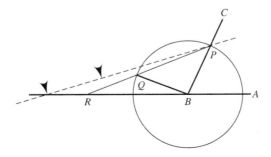

Figure 6.18

so as to attain our goal.) Now, with the straightedge passing through P, and with one mark on line AB and the other mark on the circle, we draw line PR.

We now claim that $\angle ARP = \frac{1}{3}\angle ABC$, so that we will have our desired angle trisector by constructing the line through B parallel to RP.

(6.24) PROBLEM.   In Figure 6.18, point B is the center of the circle, R lies on line AB, and secant RP is drawn, meeting the circle at Q, where QR is equal to the radius of the circle. Show that $\angle ARP = \frac{1}{3}\angle ABC$.

Solution.   Since $\angle PQB$ is an exterior angle of isosceles $\triangle RQB$, we see that $\angle PQB = \angle PRB + \angle QBR = 2\angle PRB$. But $\triangle BQP$ is also isosceles, and so $\angle RPB = \angle PQB$, and thus $\angle RPB = 2\angle PRB$. Finally, $\angle PBA$ is an exterior angle of $\triangle BPR$, and hence $\angle PBA = \angle PRB + \angle RPB = 3\angle RPB$, as desired.   ∎

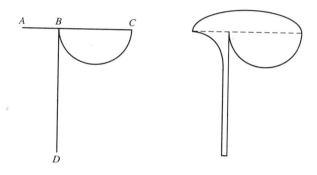

Figure 6.19

Yet another angle-trisection scheme is based on a tool that we will call a hatchet. The essential components of a hatchet are indicated in the left diagram of Figure 6.19, while an actual hatchet that can be cut from cardboard and used to trisect angles is shown on the right.

Line segment AC is divided into three equal parts, and one of the trisection points, which is not shown in the figure, is the center of the semicircle. The other trisection point is B, where the circle meets AC, as shown. Line BD is drawn perpendicular to AC, and hence it is tangent to the semicircle. We will use only point A, the semicircle, and line BD to trisect angles, but these three objects need to be held together somehow, and that is the purpose of the gadget shown on the right. The reader is urged to manufacture such a device and use it, as we are about to explain, to trisect some angles.

Suppose we want to trisect $\angle XYZ$. Place the hatchet so that point A lies on side YX of the angle, line BD passes through the vertex Y, and the semicircle is tangent to side YZ. (This is shown in Figure 6.20.) We claim that in this situation, line BD is a trisector of $\angle XYZ$.

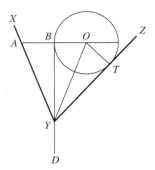

Figure 6.20

(6.25) PROBLEM. In Figure 6.20, line YZ is tangent to the circle at T and point A lies on YX. Also, O is the center of the circle, the circle meets AO at its midpoint B, and BY is perpendicular to AO. Show that $\angle XYB = \frac{1}{3}\angle XYZ$.

Solution. Draw OY and note that since BY is the perpendicular bisector of AO, we have $YA = YO$, and hence $\triangle YAB \cong \triangle YOB$. From this, we see that $\angle AYB = \angle BYO$. Furthermore, $\triangle YOB \cong \triangle YOT$ by HA since $OB = OT$, $\angle OBY = 90° = \angle OTY$, and $OY = OY$. It follows that $\angle BYO = \angle TYO$. We see now that $\angle XYB = \frac{1}{3}\angle XYZ$, as required. ∎

What happens if we modify the standard construction rules so as to make them even more restrictive than usual? Suppose, for example, that we allow the use of a compass, but not a straightedge, and suppose further that we insist that our compass be of the official classical Greek type, without a memory. What constructions are possible under these rules?

Amazingly, virtually everything that is possible with a straightedge and compass can be done with a compass alone. Of course, we cannot draw a straight line without a straightedge, but any construction that can be done with a straightedge and compass and that does not explicitly require us to draw a line can, in fact, be done with a compass alone. For example, given two points, we might be asked to construct the midpoint of the line segment they determine, or given three noncollinear points A, B, and C, we might be asked to construct the point D so that $ABCD$ is a parallelogram, or we might be asked to construct the circumcircle of $\triangle ABC$. Each of these constructions can be done with a compass alone, and more generally, every construction that is possible with a straightedge and compass and that requires us to construct a point or a circle, but not a line, can be done with a compass alone.

Although we will not present a proof of the general theorem that a straightedge is never necessary, we will give several compass-only constructions, and these will demonstrate some of the techniques that are available. We begin with a comparatively easy problem.

(6.26) PROBLEM. Given two points A and B, construct with a compass alone the point C such that B is the midpoint of AC.

Solution. Draw the circle through B centered at A and the circle through A centered at B and let P be one of the points where these circles intersect. Next, draw the circle through B centered at P and, as shown in Figure 6.21, let Q be the point other than A where this circle meets the circle centered at B. Finally, draw the circle through B centered at Q and let C be the point other than P where this circle meets the circle centered at B.

    Since points A and C each lie on the circle centered at B, we clearly have $AB = BC$, and so what remains is to show that $\angle ABC = 180°$. It is easy to see from our construction, however, that $AB = AP = BP = BC = BQ = PQ = CQ$, and thus each of $\triangle ABP$, $\triangle PBQ$, and $\triangle BQC$ is equilateral, and each of the angles of these triangles equals $60°$. It follows that $\angle ABC = \angle ABP + \angle PBQ + \angle QBC = 180°$, as required.   ∎

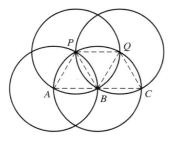

Figure 6.21

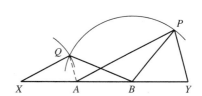

Figure 6.22

(6.27) PROBLEM. Given points A and B, and using only a compass, construct the circle centered at A and having radius equal to $\frac{1}{2}AB$.

Solution. First, use the previous problem to construct points X and Y, as shown in Figure 6.22, where A is the midpoint of XB and B is the midpoint of AY, and thus $XA = AB = BY$. Next, construct point P so that $AP = AY$ and $YP = AB$. This is easy: Take P to be a point of intersection of the circle through Y centered at A and the circle through B centered at Y. To avoid clutter in Figure 6.22, however, we have not drawn these circles.

    Now draw an arc through A centered at X and an arc through P centered at B, and let Q be the point of intersection of these arcs, as shown. That these circles really do intersect is clear from the diagram or by actually carrying out the construction. Also, we will comment later on another way to see this. We will prove that $AQ = \frac{1}{2}AB$, and thus the circle through Q centered at A solves the problem. To avoid clutter in Figure 6.22, we have not drawn this circle either.

    First, we argue that $\triangle XQB \cong \triangle ABP$. We have $XQ = XA = AB$ and $XB = 2AB = AY = AP$. Also $BQ = BP$, and thus the triangles are congruent by

SSS. It follows that $\angle QXB = \angle BAP$, and we can rewrite this as $\angle QXA = \angle PAY$. Now $AY = 2AB = 2XA$ and $AP = AY = 2XA = 2XQ$, and thus $AY/XA = 2 = AP/XQ$. By the SAS similarity criterion, this proves that $\triangle QXA \sim \triangle PAY$, and thus the ratio YP/AQ is also equal to 2, and we have $AQ = \frac{1}{2}YP$. But $YP = AB$, and thus $AQ = \frac{1}{2}AB$, as wanted.   ∎

It is not hard to calculate the distance BP in Figure 6.22, and this provides an alternative argument to show why the two circles whose intersection defines the point Q really do intersect. For simplicity, let us suppose that $AB = 1$. Then also $YP = BY = 1$ and $AP = 2$. By Stewart's theorem (Theorem 2.20) applied in $\triangle APY$, we have $4 + 1 = 2t^2 + 2$, where $t = BP$. It follows that $BP = t = \sqrt{3/2}$, and thus $BA = 1 < BP < 2 = BX$. Thus A lies inside the circle through P centered at B, and X lies outside of this circle. It is now clear that the circle through A centered at X must meet the circle through P centered at B, as we saw in our construction.

We can now construct the midpoint of the line segment determined by two given points.

(6.28) PROBLEM.   Given points A and B, construct with a compass alone the midpoint of line segment AB.

In some sense, this is now very easy. Just use Problem 6.27 twice to construct circles centered about each of A and B and having radii equal to $\frac{1}{2}AB$. These two circles will be tangent, of course, and the unique point where they meet will be the desired midpoint. This seems to be cheating a little, however, and so instead of worrying about whether or not we have described legal construction, we present an alternative whose legality should not be in doubt.

Solution to Problem 6.28.   Let P and Q be the two points of intersection of the circle through A centered at B and the circle through B centered at A. Then $APBQ$ is a parallelogram, and so the point we seek, the midpoint of AB, is also the midpoint of PQ, and hence it lies on the circle centered at A with radius $\frac{1}{2}AB$ and also on the circle centered at P with radius $\frac{1}{2}PQ$. Each of these circles can be constructed by Problem 6.27, and we see that one of the two points of intersection of these circles is the desired midpoint. If there is any doubt about which of these two intersection points is the correct one, draw the circle centered at B with radius $\frac{1}{2}AB$. That circle will pass through just one of the two intersection points, and that is the point we want.   ∎

It is now easy to do the complete-the-parallelogram problem, mentioned earlier.

(6.29) PROBLEM.   Given three noncollinear points A, B, and C, construct with a compass alone the point D such that $ABCD$ is a parallelogram.

Solution.   First, use Problem 6.28 to construct the midpoint M of AC and then use Problem 6.26 to construct the point D such that M is the midpoint of BD. Then

diagonals AC and BD of quadrilateral $ABCD$ have the common midpoint M, and so these diagonals bisect each other. It follows by Theorem 1.9 that $ABCD$ is a parallelogram, as desired.
∎

It is a consequence of Problem 6.29 that given three points A, B, and C, and working with a compass alone, we can draw a circle of radius AB centered at C. To do this, construct D as in Problem 6.29 and draw the circle through D and centered at C. Even for compass-alone constructions, therefore, we can assume that our compass has a memory.

We close with a problem that is an example of another variation on the theme of straightedge and compass constructions. Since the ideal Greek construction tools are allowed to use the entire Euclidean plane, we have not, up to now, considered the size or shape of the paper on which we are working. But consider this problem.

(6.30) PROBLEM.   Given nonparallel line segments AB and CD on a rectangular piece of paper and a point P lying between them, let X be the point where lines AB and CD meet, but do not assume that X lies on the paper. Construct a segment of the line PX, using only a straightedge and compass, but keep all of the construction work on the paper.

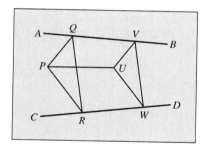

Figure 6.23

Solution.   The enclosing rectangle in Figure 6.23 represents the paper on which we are working. Choose arbitrary points Q and V on AB and R on CD, as shown. Draw lines QP and QR and construct lines through V parallel to these two lines. In particular, this determines the point W on CD such that $VW \| QR$. Next draw RP and construct the line through W parallel to RP. As shown in Figure 6.23, this determines a point U such that $WU \| RP$ and $VU \| QP$. Finally, draw PU, which is the desired line segment. The fact that line PU passes through X is a consequence of Theorem 6.31, which we state and prove separately.
∎

(6.31) THEOREM.   *Suppose that corresponding sides of $\triangle PQR$ and $\triangle UVW$ are parallel. Then lines QV, PU, and RW are either concurrent or parallel.*

Proof.   If the three lines are parallel, there is nothing to prove, and so we assume that two of them (say, QV and RW) meet at some point, and we will use vectors to show that point also lies on PU.

First, since $PQ \| UV$, we can write $\overrightarrow{UV} = \alpha \overrightarrow{PQ}$ for some scalar α, and similarly, $\overrightarrow{UW} = \beta \overrightarrow{PR}$ and $\overrightarrow{VW} = \gamma \overrightarrow{QR}$ for some scalars β and γ. Since $\overrightarrow{UW} = \overrightarrow{UV} + \overrightarrow{VW}$ and $\overrightarrow{PR} = \overrightarrow{PQ} + \overrightarrow{QR}$, we have

$$\beta\left(\overrightarrow{PQ} + \overrightarrow{QR}\right) = \beta \overrightarrow{PR} = \overrightarrow{UW} = \overrightarrow{UV} + \overrightarrow{VW} = \alpha \overrightarrow{PQ} + \gamma \overrightarrow{QR},$$

and thus $(\beta - \alpha)\overrightarrow{PQ} = (\gamma - \beta)\overrightarrow{QR}$. But PQ is not parallel to QR, and so the only way that a scalar multiple of the vector $\overrightarrow{PQ}$ can equal a scalar multiple of the vector $\overrightarrow{QR}$ is if both scalars are 0. In particular, we conclude that $\beta - \alpha = 0 = \gamma - \beta$, and so $\gamma = \beta = \alpha$, and we have $\overrightarrow{UW} = \alpha \overrightarrow{PR}$ and $\overrightarrow{VW} = \alpha \overrightarrow{QR}$. In particular, if $\alpha = 1$, we see that $\overrightarrow{QR} = \overrightarrow{VW}$, and thus $QVWR$ is a parallelogram. This is not the case, however, since we assumed that lines QV and RW have a common point, and we conclude that $\alpha \neq 1$.

We choose the point P to be our origin so that by the notational convention of Chapter 5, we can write $\vec{Q} = \overrightarrow{PQ}$ and $\vec{R} = \overrightarrow{PR}$. Also,

$$\vec{V} = \vec{U} + \overrightarrow{UV} = \vec{U} + \alpha \overrightarrow{PQ} = \vec{U} + \alpha \vec{Q}$$

and similarly, since $\beta = \alpha$, we have

$$\vec{W} = \vec{U} + \overrightarrow{UW} = \vec{U} + \alpha \overrightarrow{PR} = \vec{U} + \alpha \vec{R}.$$

Since $\alpha \neq 1$, we can write $\lambda = 1/(1 - \alpha)$, and we consider the point X on line QV such that $\overrightarrow{QX} = \lambda \overrightarrow{QV}$. We have

$$
\begin{aligned}
\vec{X} &= \vec{Q} + \overrightarrow{QX} \\
&= \vec{Q} + \lambda \overrightarrow{QV} \\
&= \vec{Q} + \lambda(\vec{V} - \vec{Q}) \\
&= \vec{Q} + \lambda(\vec{U} + \alpha\vec{Q} - \vec{Q}) \\
&= \lambda \vec{U},
\end{aligned}
$$

where the last equality follows since $1 + \lambda\alpha - \lambda = 0$. Similarly, if Y is the point on line RW such that $\overrightarrow{RY} = \lambda \overrightarrow{RW}$, we compute that $\vec{Y} = \lambda \vec{U}$. It follows that $\vec{X} = \vec{Y}$, and thus X and Y are the same point, and we conclude that X is the intersection point of lines QV and RW.

We know that $\overrightarrow{PX} = \vec{X} = \lambda \vec{U} = \lambda \overrightarrow{PU}$, and this shows that X lies on line PU. Since X also lies on both QV and RW, it follows that lines PU, QV, and RW are concurrent, as required.   ∎

We mention that Theorem 6.31 can be thought of as a special case of the converse of Desargues' theorem in projective geometry. (See Exercise 4D.7.)

Exercises 6F

6F.1 Suppose that our only construction tool is a two-edged straightedge, where the two edges are parallel. Show how to construct the perpendicular bisector of a given line segment AB.

HINT: Do the case first where the length of the given segment exceeds the distance between the two edges of our straightedge. Begin by placing the straightedge on the paper so that one of its parallel edges runs through A and the other edge runs through B.

6F.2 Using only a two-edged straightedge, erect a perpendicular to a given line through a given point on the line.

6F.3 Using only a two-edged straightedge, construct a line parallel to a given line through a given point.

HINT: Begin by drawing three parallel equally spaced lines, where one of these lines goes through the given point. Consider the points where these three lines meet the given line.

6F.4 Using only a two-edged straightedge, drop a perpendicular to a given line from a given point not on the line.

6F.5 Using only a two-edged straightedge, construct a regular octagon.

6F.6 Suppose that we work on lined paper, where the printed lines are parallel and equally spaced. Using only an ordinary straightedge, construct a line through a given point parallel to a given line.

HINT: This problem is much easier if the given point lies on one of the printed lines. Do that case first.

6F.7 Given points A and B and using only a compass, construct the circle centered at A with radius $\frac{1}{3}AB$.

Some Further Reading

Needless to say, for a subject as old as geometry, there are countless books on the subject dating from Euclid's *Elements* up to the present. The purpose of this brief bibliography is to introduce a few works that we believe might be of interest to readers of this book. We begin, of course, with Euclid, himself.

- Euclid, *The Thirteen Books of the Elements* (3 vols.), translated with an introduction and commentary by Sir Thomas L. Heath. Dover, New York, 1956.

This is probably not the best source if you really want to learn more geometry, but it is wonderful to see the book that started it all. In these volumes, one can also find an extensive and scholarly historical commentary. Heath's translation was first published in 1908 and a revised version appeared in 1925.

There is vastly more to geometry than we have discussed or even mentioned in this book. A sampling of some of this other material can be found in the following work by one of the most distinguished geometers of modern times.

- H. S. M. Coxeter, *Introduction to Geometry*. Wiley, New York, 1961.

In the first 25 or so pages of Coxeter's beautiful book, readers will find a number of familiar topics and theorems: the Euler line, the nine-point circle, Fermat's minimization problem, and Morley's theorem, to name some of them. But Coxeter goes on to discuss completely different geometric ideas. These include symmetry groups, projective geometry, non-Euclidean geometry, differential geometry, topology, and higher dimensional geometry. Coxeter's book should not be thought of as a compendium of theorems; it is a bouquet of ideas, somewhat loosely tied together by the theme of geometry.

As Coxeter's *Introduction to Geometry* demonstrates, there is much more to geometry than "yet another theorem." Nevertheless, it is also true that there are many beautiful theorems that we have not been able to include in this book. Some of these can found in the following work, which seems well on its way to becoming a classic.

- H. S. M. Coxeter and S. L. Greitzer, *Geometry Revisited*. New Mathematical Library, Mathematical Association of America, Washington, D.C., 1967.

Geometry Revisited includes many of the topics we have discussed and a number that we have not. It also includes some interesting historical commentary and a large number of exercises with hints and answers. The large overlap in topics between *Geometry*

Revisited and this text is not a coincidence since, in fact, I used Coxeter and Greitzer as a source of ideas for the geometry course upon which this book was based.

Still more really neat theorems can be found in the following work.

· R. Honsberger, *Episodes in Nineteenth and Twentieth Century Euclidean Geometry.* New Mathematical Library, Mathematical Association of America, Washington, D.C., 1995.

As the title suggests, the focus of this book is on some of the newer discoveries in geometry. Among these, we cannot resist mentioning an especially pretty result about isogonal conjugates: If P and Q are isogonal conjugates in the interior of $\triangle ABC$ and perpendiculars are dropped from each of these points to the sides of the triangle, then the six feet of these perpendiculars lie on a circle. The proofs in this book are clearly written and easy to follow, although not all of the amazing theorems that the author mentions are proved.

Here is the answer for readers who feel that we did not provide enough, or hard enough, exercises in this book.

· A. S. Posamentier and C. T. Salkind, *Challenging Problems in Geometry.* Dover, New York, 1996.

In this reprint of a book first published in 1970, there are solutions as well as problems.

Finally, we come to what can only be described as an encyclopedic collection of theorems in Euclidean geometry.

· N. Altshiller-Court, *College Geometry.* Barnes and Noble, New York, 1952.

This book, which is a reprint of a work first published in 1925, contains a vast number of interesting theorems and an equally vast collection of exercises. Although this book was written as a text, many of today's college students might find Altshiller-Court's proofs somewhat difficult to follow. Nevertheless, this is the place to go to look up some obscure geometric fact. Unfortunately, this book has long been out of print, but it should be available in a good library.

Index